P9-DWG-821

D0037596

Pehr G Gyllenhammar
President **VOLVO**

# People at Work

**ADDISON-WESLEY PUBLISHING COMPANY**
Reading, Massachusetts · Menlo Park, California
London · Amsterdam
Don Mills, Ontario · Sydney

To those who disagree with me and those who
educated me. They are sometimes identical and
always include my wife Christina.

# Introduction

I am pleased to have the opportunity to write this introduction to Pehr Gyllenhammar's important contribution, *People at Work*. I first visited Volvo's Torslanda plant in August 1973, and the Kalmar facility in July 1974. The evidence was clear that significant contributions to the restructuring of industrial life were under way.

In the period immediately prior to these years, there had been much discussion in the United States about the infinite boredom of factory work and many treatises written about the "blue-collar blues." Much of this discussion was patronizing and intensified the suspicion in workers' minds that they were going to be used.

It is interesting that Pehr Gyllenhammar writes: "We talk a lot about blue-collar work and its problems, but I think white-collar work is the most deadly dull today."

The first Henry Ford's moving assembly line has brought much good to the world. Through creating a mass-production system, it has fostered a mass-consumption system.

It also brought much that was wrong. The assembly line eventually led to an oppressive serf system in the Ford Motor

Company itself, lorded over by Harry Bennett. Workers on the line were treated as automatons, as subhumans.

Much has changed and is changing. Managers are recognizing that the human element is paramount. In this book, all involved in the industrial process are recognized as "stakeholders," and, beyond that, as human beings with a priority of interest.

It is refreshing that a growing cadre of business managers, of which Mr. Gyllenhammar is representative, is introducing into the private-enterprise system a philosophy based on humanism —recognition of the dignity and self-respect of the individual. The humanistic approach guides business decisions in directions heretofore largely neglected, yet vital to society's progress if we are to achieve the brave new world.

Under the system of private enterprise, the essential purpose of business is not so much to produce products or perform service, but rather simply to make profit. And, of course, no private enterprise can long endure without being profitable. Pehr Gyllenhammar's view of the purpose of business, springing from his humanistic concepts, expands far beyond the bounds of sheer profit making. "What is the purpose of business?" he asks. And he responds, "I believe the purpose of private enterprise is to serve the public."

Guided by this fundamental philosophic underpinning, his concepts concerning a multitude of facets of work and of business life take shape in a way that departs from traditional business thought and practice.

The idea of reassessing old habits and implementing change to meet the modern condition is the natural outgrowth of this philosophy. One divines early that, in his view, change must not be forced. Rather, change should evolve naturally, because socially, morally, and philosophically it makes sense. Yet, he never loses sight of the need to achieve the successful outcome. The bottom line on the balance sheet must be written in black ink. It is precisely this synthesis of sound moral and human motivation with good business management that makes Pehr Gyllenhammar's statement both pertinent and intriguing.

In an industrial world believing firmly in the economy of

scale—bigger is beautiful—Mr. Gyllenhammar argues that there is a limit to the necessity and efficiency of large-sized factories. In a system relying increasingly on technology, he asserts that the machine is, after all, the creation of the human being, and not vice versa. The conclusion to be drawn, therefore, is that work must be reorganized to suit people, and this in turn requires changing the technology that chains people to the machine. "It is cruel . . . to expect people to spend more than half their working hours each day . . . simply acting as machine tenders," he writes. Or, as he notes in another passage: "Creating educated automatons is unacceptable if you view people as adults who can develop in a number of directions."

His philosophy has been practically applied in both the older and newer facilities of Volvo. No master plan guides the direction and tempo of change; but, within the limitations of transferability, and taking into account the varied customs, habits, and traditions of relationships in each situation, earnest effort is made to tailor each innovation to the given circumstances.

Reorganizing work to involve workers more meaningfully in the decision-making process and to create jobs that are "human and meaningful" is the central theme of Mr. Gyllenhammar's writing. But his concepts reach out beyond the workplace to the community and even the world at large. The humanistic approach to business administration is captured in his statement: "The most important guideline would be for a company never to contribute deliberately to activities that are contrary to fundamental principles concerning human rights."

The company therefore must "administer the resources with which [it] is entrusted . . . to create economic growth, taking into consideration all the interest groups involved with the company." And this includes consideration not only of the stockholders and the managers, but the customers, the suppliers, the employees, the government, and society as a whole. It requires a code of ethical behavior for business executives to abide by—both in their domestic operations and their multinational operations. There is need, therefore, for self-examination, a review of business values to keep up with changes in public attitudes. Businesses must be willing to provide the public with objective summaries of technical information available; only then can the public properly evaluate progress achieved and projected.

Thus, Mr. Gyllenhammar advocates the use of a "social audit" of the company's operations, that is, "taking the outside into account." The concept of the social audit reflects back to the philosophic precept that the purpose of business "is to help achieve and maintain the public good."

Pehr Gyllenhammar has set as his objective the fulfillment of noble democratic values, not only in society in general, but in the workplace as well. He wants to achieve "democracy in working life"—and that is the next step toward the consummation of the truly democratic way of life.

In realistic terms, he describes imaginative, creative concepts of work organization and the new technology that has been implemented successfully in fulfillment of these stated objectives. While this small volume does not describe in any direct way worker reaction to these new managerial directions, Mr. Gyllenhammar depicts the success of the various programs in some detail. For example, regarding the program at the plant in Olofström, he notes:

- *Employee turnover was cut to one-quarter the previous average.*
- *Absenteeism dropped to one-half the previous average.*
- *Recruitment of new employees became easier.*
- *Quality of product improved.*
- *Labor productivity was maintained.*

It is equally important to note that "profitability was maintained."

Fundamental to Pehr Gyllenhammar's philosophy of good management is the expressed belief that each individual has a right to have a job; has a right to self-respect in his or her work; has a right to "good work," work that is "human" and "meaningful," in a circumstance that permits workers to have "as much control as possible over their working lives"; and has a right to "decent pay."

The question is: Does it work? The answer has to be, yes. And not just in Sweden, which is remarkable for its highly sophisticated and mature labor movement, both blue-collar and white-collar.

This book does not offer a ready blueprint. As Pehr Gyllenhammar writes: "There are no experiments. There is no single solution."

The magic is the identification of "stakeholders" and free human beings.

Beyond doubt, traditionalists in management will look on the ideas expressed in this volume with a jaundiced eye. But Pehr Gyllenhammar has stirred the winds of change. He represents the challenge of the present and the future in business management and in labor-management relations.

*March 1977*                    LEONARD WOODCOCK
                               *President, International Union,*
                               *United Automobile, Aerospace,*
                               *and Agricultural Implement*
                               *Workers of America (UAW)*

# Preface

In 1969, I wrote my first book and called it, *Towards the Turn of the Century, At Random*. Its purpose was to express concern about the way our politicians handle our future. In democratic systems politicians are elected for a limited period, so one of their natural preoccupations is ensuring their own reelection. But as our societies grow more complex, problems cannot be solved unless one takes a long-term view. The politicians tend to give top priority to the short-term issues. Therefore, I expressed concern that problems crucial to humankind are being neglected.

In 1973 my second book, *I Believe in Sweden,* was published. I expressed opposition to those who saw government intervention as generally detrimental to society and suggested that the Swedish system was well designed and functioned smoothly because of good "engineering" and because our society was less complex than others. The country's small and homogeneous population was an advantage. My own attitude towards the Swedish system was positive, but I expressed concern that it had become so elaborate that too little was being left to the individual, the major driving force behind the country's healthy development. I could also see tendencies toward arrogance in the way our public administration viewed people's demands. Democratic societies in the industrialized world are growing, not only in size but also in complexity. Along with better living standards and education come higher aspirations for the individual. These aspirations impose certain restrictions on the activities of our institutions, including corporations. This makes management a more delicate job. New stakeholders arise alongside the traditional ones. To maintain a balance between all the different stakeholders has thus become one of the major responsibilities of any manager or business leader in today's world. Like others in my position, I have had to give a great deal of thought to the preconditions for corporate life and how one can possibly solve the increasing number of potential conflicts. Because it might be useful to share my experiences with others, I have brought them together in this book.

Most people find it easier to deal with such subjects as new organizational structures and production schemes in abstract terms rather than in concrete descriptions. The obvious reason is that these things have been discussed a great deal, but few have actually been put into practice. Because I have seen too few examples in the literature of work in this area, I have structured my book to restore some balance to the discussion of corporate development. Even so, the number of pages dealing with production methods and organization development should not be interpreted as meaning that I consider this area of corporate activity to have overriding importance.

One group of stakeholders whose interests have become more prominent in recent years are the employees. Because of this development, and because I have found that human resources are our most important asset, it is natural that people should play a key role in this book.

## ACKNOWLEDGMENTS

My Swedish publisher Timo Kärnekull of Askild & Kärnekull tempted me and encouraged me to write a book for Swedish readers in 1973. He is one of the few people I know who could persuade me to do something that I originally did not want to do. His American friend and colleague, Warren Stone, is another one of the few. He persuaded me to write a new book for American readers. I am deeply grateful for Warren's encouragement.

I have a great deal to do, without using my time to write books, although I do write anyway as a stimulus during my free time. Nancy Foy, a freelance journalist, is a close friend who has taken a deep interest in some of my work. When I asked Nancy for advice on my writing a book, she offered her assistance. Only for that reason did I decide to go ahead. Nancy has collected, studied, and evaluated all my fragmented material and helped me to make a book out of it. Mutual understanding and friendship have been the basis of the task. What may not be good is my fault and what is acceptable is to Nancy's credit. The thinking is something for which I take all the responsibility.

My associates Margareta Eriksson and Gun Fornmark manage me anyway. Getting this book off the ground has been no exception.

Two of my colleagues see it as their job to criticize my job. This is fine, and good for survival. So I gave the manuscript to Bo Ekman and Berth Jönsson. Their advice has been very useful.

My wife Christina still thinks that I am talented and that I have my future made, although she does not see this book as any indication thereof. She has inspired me to write some essential revisions.

*Gothenburg, Sweden*                                            PGG
*March 1977*

# Contents

# 1

# A working philosophy

**P** eople, not machines, are the real basis for the spectacular growth of industry during the twentieth century. Yet when we talk about the technological world today, we tend to give credit to the products, not the people:

- *"The march of technology has created systems of unbelievable complexity . . ."*
- *"The computer made a mistake in my electric bill . . ."*
- *"The automobile has changed the way we live . . ."*
- *"He owes his life to the kidney machine . . ."*
- *"Multinationals are not accountable to government . . ."*

We hear statements like these every day. But when is mention made of the many people involved in these advances? Each technologist contributed a little in his or her own specialty, and the sum of all their efforts, responding to our needs, was the march of technology. The systems of unbelievable complexity are the creation not of technology but of individual system designers. It was not the computer that made the mistake in an electricity bill but the designer, the programmer, or perhaps the person who read the meter or typed in the reading. The same designer or programmer might also be responsible for the computerized intensive-care unit in a hospital. The people who made the kidney machine and the medical technician who runs it are the ones who should get credit for life saving, not the machine itself.

Similarly, the automotive age didn't just spring upon us out of nowhere. Today's car is the product of designers, car builders, production engineers, dealers, and many more, responding to what they believe are the demands of customers. The huge companies we know as multinationals are also made up of individuals. They have grown up from small ventures, started by individual entrepreneurs. Later generations of marketing specialists, economists, organization developers, accountants, production experts, and managers welded them together into the form we have today.

I am concerned that the success of the technology, the systems, the organizations, and the companies have begun to obscure the objectives their creators had in mind: improving the lives of individuals grouped together in an even larger unit we call "society."

I am also concerned that developing countries, admiring the economic success of the industrial countries of America and Western Europe, will emulate our present patterns without exploring these underlying objectives. Thus, they may find themselves with all our present problems at a time when we ourselves are beginning to question the economies-of-scale rationale.

## THE INDUSTRIAL ILLNESS

In any industrialized country today more than half the working population still works in companies with less than fifty employees. But some giant organizations are growing even larger—too large to be comfortable for employees. The individual working in a large modern company too often feels lost in the overall scheme, merely a replaceable cog in the industrial system, with little or no control over his or her own life until retirement. The same anonymity occurs inside government, as departments grow to cope with social problems and the growth itself creates new social problems. In some countries the government employs at least one-third of the working population, directly or indirectly.

Much of our growth resulted from the discovery that a larger unit could get access to resources—notably capital—that were unavailable to smaller units. We found ways to produce in large factories more efficiently than we could in small workshops where a small team of workers was the basic manufacturing unit. We found ways to move our goods to more distant markets. We began to move people as well, creating new demands for products and an international flow of attitudes and values, as well as what we call "technology transfer." In these circumstances, it is not surprising that enterprising individuals often decided that if a little bigness was good, a lot of it must be better. Their decisions gave rise to such phenomena as conglomerates, thousand-desk offices, and mile-long assembly lines.

Like other good things, economy of scale turned out to have subtle limits. We begin to find today the symptoms of a new type of industrial illness. We invent machines to eliminate some of the physical stress of work, and then find psychological stress causing even more health and behavior problems. People don't want to be subservient to machines and systems. They react to inhuman working conditions in very human ways: by job-hopping, absenteeism, apathetic attitudes, antagonism, and even malicious mischief. From the worker's point of view, this is perfectly reasonable. The younger the worker is, the stronger his or her reactions are likely to be. People entering the workforce today have received more education than ever before in history. We have educated them to regard themselves as mature adults, capable of making their own choices. Then we offer them virtually no choice in our overorganized industrial units. For eight hours a day they are regarded as children, ciphers, or potential problems and managed or controlled accordingly.

Today a new element enters the picture—permanent unemployment. Our technical advance has been so rapid that we can produce more with less people, just at a time when more people are entering the job market. Workers whose jobs are threatened in the name of "progress" cluster together to protect the status quo. The featherbedding situation that follows not only hastens the demise of their organizations but also creates polarized "them and us" attitudes that make it impossible to propose retraining, work reorganization, and other approaches that might help create new jobs and new business. The more workers try to protect their endangered jobs, the more planners and managers simply see their attitude as reason to design people out of the system.

If the industrial society, with its intensive technological development, succeeds in putting people to one side, what will happen to them and, thus, to society itself? What kind of work will be available in the future as an alternative to the jobs that industry offers today?

Humane answers to these crucial questions depend on the cultural and economic framework in which industry operates—and on the commitment of the people making the decisions that will affect employment in the future. Instead of doing away with jobs, they will have to ask themselves how to make jobs bearable, let alone satisfying and meaningful. We need to find ways to personally involve each worker. Management cannot be based on power. In any show of power, the workers will "win" and management will "lose," although the real result is, inevitably, that everyone loses.

## THE SWEDISH WAY IS BEST—FOR SWEDEN

For reasons of its own, Sweden has spent a great deal of time and energy in the past few years looking at some of these problems and trying to find ways, in law and in practice, to make work more liveable and life more workable. We have stringent new laws for job security and safety at work. A 1977 law calls for full consultation with employees and full participation by employee representatives in a wide range of decision making, from board-level to shop-floor matters. Union agreements give employee organizations the right to all financial information about a company. More important than the laws are the attitudes of managers, who seem to share the conviction that greater participation is good and necessary in spite of the discomforts of change.

Sweden is in a unique position, with a population less than the city of Los Angeles spread over an area larger than the state of California. The country is more dependent on its success in world markets than many larger countries. The population is fairly homogeneous, mainly Nordic, and this gives rise to a coherent culture in which social change can be effected more swiftly than it can in places like Britain or the United States which have more heterogeneous populations and many different cultures within their borders.

Industrially, Sweden's situation is quite different from, say, that of the United States. While the United States has a pool of unemployed that, in effect, masks manpower problems from management, Sweden has relatively full employment, and is to some extent dependent on "guest workers" from such countries as Finland and Yugoslavia. Sweden and the United States, though, have a similar need today to involve people in their work in order to achieve better productivity. Sweden has its labor laws, a body of practice, a union/employer structure, and an entire industrial culture based on cooperative rather than contentious models. Blue-collar union membership is around 90 percent, and wage agreements are worked out from a national perspective.

The United States model is based more on arm's length negotiations, competing unions, and much lower union membership. However, both countries have fairly streamlined union organizations in any specific factory, and both have clear-cut ways of communicating with the unions, with relatively strike-free results overall.

One of the problems both countries share is employee absenteeism, but a major difference between the two is that Sweden experienced the problem sooner and has treated it as a more serious matter. Many United States companies, on the other hand, consider it unimportant in view of the millions who wait hopefully for jobs. This difference deserves attention because I believe it is fundamental to understanding the difference between Swedish and American attitudes towards employee participation.

Employee turnover problems in Sweden became noticeable towards the end of the sixties, at a time when there were other indications of social change as well. Compulsory education had recently been raised from seven to nine years. The student uproar that prevailed in other parts of the world surfaced in Swedish universities as well. Such issues as preserving the environment and improving working conditions began to gain more public attention. People became more mobile in their employment than they had ever been before. Employee turnover peaked in 1970 at up to 50 percent in certain metropolitan industries. This was the exception, not the norm, but it was a strong indication that things were going wrong. High mobility not only causes problems of recruiting and training, it also indicates a growing disaffection with work. People tend to go on strike or express grievances about matters that have to do with money, but the things that really

frustrate them at work tend to concern working conditions of a subtler sort. You don't go on strike if you get bored with your job. You just leave the job, or give yourself a holiday. The advances Sweden had made in social benefits meant that workers could do this with very little financial penalty; our unemployment and sick-pay benefits are among the highest in the world. When our problems were at their peak, I visited a few colleagues in the United States and asked them about job-hopping. "No problem," was the usual answer. I was amazed, especially after I had visited some United States plants. Their work patterns were pretty similar to ours in Sweden at the time—turning out the same products in the same assembly lines, day after day, year after year, using the same procedures, being measured and controlled by the same sophisticated systems. It was clear that the American workers had the same problems our workers had. The main difference was that the American managers didn't perceive it as a serious problem. America was accustomed to greater mobility anyway, and, after all, managers reasoned, they could train people and put them on simple jobs so quickly that the turnover figures didn't matter, although planning was a nuisance for plant managers.

A closer investigation showed that some American plants I visited had turnover figures similar to Sweden's—which meant that, at the worst, half the employees left every year. To me this seemed horrifying—that the average worker would feel the need to change jobs at least every two years. I'm still convinced that boredom is a major problem. The conventional system carries within itself the seeds of its own collapse, even though the different labor availability changes the perspective for managers outside Sweden.

Management attitudes notwithstanding, it is a basic human fact that if you raise people's expectations and then disappoint them often enough, for long enough, they will grow angry. And this is what we are doing. By its nature, routine factory work has no correlation with what we seem to offer people in the educational system. Until they are eighteen or so, students are encouraged to learn about ideas, sciences, history, literature, languages, art, and music and to develop some job related skills as well. Many go on to university with the idea that it will improve their job chances. Then from eighteen, or twenty-two, to sixty-five, we expect them to do exactly the same thing for eight hours a day, minus coffee breaks, sick leave, holidays, and strikes. Creating educated automatons is unacceptable if you view people as adults who can develop in a number of directions—as human beings with enormous potential. Given this view, which I hold strongly, it is cruel to the individuals and wasteful to society to expect people to spend more than half their waking hours each day without stimulus of any sort, simply acting as efficient machine-tenders. In spite of the differences between Sweden's culture and that of other industrial countries, to a large extent governments share the basic premise that their citizens are adults. Thus, to some extent, other countries will have to cope with the same problems.

## THE VOLVO EXPERIENCE

Just as Sweden's situation differs from that of other countries, so does Volvo's differ from other companies. The events at Volvo's factories need to be seen in that perspective.

Volvo is by international standards a medium-sized company today, but in its home country, Sweden, it is a large company. Ranking between sixtieth and seventieth in the list of non–United States companies, Volvo's share of the world automotive market is small, just over 1 percent. However, the company derives 70 percent of its sales from outside Sweden and accounts for between 8 and 9 percent of Sweden's total export.

When they started the company in 1927 in Gothenburg, Sweden, Volvo's founding fathers assumed that Sweden's engineering capability would help the company take a share of the rapidly growing market for automotive transportation. They based their company on a strong commitment to quality. In spite of limited exports, their venture grew as Sweden developed into an industrialized country.

Armed with a workforce that was stable, competent, and hard working, Volvo entered a new phase in the late fifties by increasing its export potential dramatically. The company decided to enter the important United States market, and also established spearheads in Europe. In the fifties, cars accounted for more than 70 percent of sales and 80 percent of profits, but the product line grew to include trucks, buses, industrial machines, farm equipment, marine engines, and jet turbines for the Swedish Air Force. By the late sixties, the company had become an internationally oriented organization, with new subsidiaries in Europe and beyond, and important manufacturing facilities within the Common Market, notably in Belgium.

Throughout the sixties, the Volvo organization was tightly centralized, controlled by the president in a three-man executive committee, assisted by a large head office staff. Then, in 1969, the company introduced a more decentralized structure to give more stature and resources to products other than passenger cars. Subsidiary companies outside Sweden continued to report to the member of the executive committee responsible for marketing.

Like other auto manufacturers, Volvo had a production system that was technically oriented and planned in detail, using the system of Methods Time Measurement (MTM). The labor unrest that became visible in 1969 made it necessary to adapt production control to changing attitudes in the work force.

To increase export and build further volume and profit in the non-car divisions, Volvo went through a major reorganization in 1972. Headquarters was cut from 1800 to 100 people; most of the former headquarters people went into profit centers offering consulting and services. Each major product group became an independent division, and all major market units became independent profit centers within the divisions. The new organization took on a much more international flavor, recognizing the fact that more than 100 countries provided its income. To balance the income from the dominant car activities, major investments were made in trucks, buses, and industrial equipment.

The oil crisis of 1973 and the recession that followed proved the value of the reorganization. The company emerged with a reduced but still healthy profit, and, for the first time, the majority of the profit came from non-car operations.

Today Volvo is still very much a Sweden-based company and we intend to keep it that way, although investments outside the home country will become more prominent in our plans. Today more than 80 percent of Volvo's total assets in plant and equipment are spread over twenty-seven communities in Sweden. However, one-third of new investment is international. A major 1975 acquisition was DAF Car BV in the Netherlands, now named Volvo Car BV. Through this Dutch affiliation, Volvo has been able to widen its narrow range of cars.

Although Sweden's labor costs have become the highest in the world, Volvo remains a profitable company. In 1974 and 1975 the company's cash flow as a percentage of sales was one of the three highest in the automotive industry. The corporate strategy today is to add strength and reduce vulnerability by diversifying within the field of transportation.

Chapters 3–5 describe some of the changes Volvo has made over the past few years to improve conditions for employees. We don't think of the new factories at Kalmar and Skövde, or the changes in our existing factories, as "experiments." The word "experiment" implies uncertainty, a lack of determination to go ahead, and an unwillingness to make a strong commitment. There is no time today to deal with huge problems on a miniature scale. We have to deal with them full scale, whether they are problems in society or problems in working conditions.

The important thing to remember is that there is no way back from major changes in working patterns. The Kalmar plant, for example, is designed for a specific purpose: car assembly in working groups of about twenty. If it didn't work, it would be a costly and visible failure, both in financial and social terms. In that case, we would lose credibility, with our own people as well as with those who were watching gimlet-eyed from outside. Fortunately, it worked—quite effectively. Now we have not one but five new plants, organized in a nontraditional way, all scaled for 600 employees or less. These cost a little more to build than traditional factories of similar size, but they are already showing good productivity. We believe productivity will continue to increase because the people who work in them have better jobs. The technical improvements are already showing results. We hope the employees will continue to be motivated to give more of themselves, their ideas, and their creative energy to the endeavor.

When we started thinking about reorganizing the way we worked, the bottleneck seemed to be production technology. You can make a corporation spend almost any amount of money these days on marketing or sales promotion without any evidence at all that the investment will buy them anything. If you asked a large corporation to cut down its advertising budgets, top executives would feel uneasy. Perhaps the only way a company could measure the real effect of advertising would be to stop doing it—but nobody dares, because some executive would be blamed if the effects were adverse. In the same vein, traditional reasoning has been that the present technology has a more than credible image throughout its history, ever since Henry Ford created the assembly line for mass production of complex mechanical products. If the present technology works, then why would a company change it for something risky, something untried, something new?

We found what we felt were good reasons for change. Acting in the belief that we couldn't really reorganize the work to suit the people unless we also changed the technology that chained people to the assembly line, we took some steps that seemed risky at the time—especially because they were irreversible. In a new factory we broke up the inexorable line to which the workers were subservient, and replaced it with individual carriers that move under control of the workers.

An assembly line is essentially a set of conveyors going through a warehouse full of materials. The materials are the focus of the system, not the employees. People are constantly having to run after their work as it moves past their stations.

We started with the idea that perhaps people could do a better job if the product stood still and they could work on it, concentrating on their work, rather than running after it and worrying that it would get beyond them. So we developed the industrial carriers, each one carrying a single product. At Kalmar the product is a car; at Skövde the product is an engine, but the principle remains: the carriers can run around under their own power, in layouts that can be changed easily. The carriers move according to the desires of the workers now, instead of the workers moving to keep up with the line.

Another problem in factory life was the antisocial atmosphere built into the production line. People want to have some

social contact. But in an assembly-line situation, even if the plant is quiet, people are physically isolated from each other. In the noise of the traditional auto plant, people typically have to yell over the sound of machines if they do manage to get together to discuss something. Furthermore, workers are distracted by jobs of such short duration (perhaps thirty to sixty seconds) that they seldom get a chance to stop and think or talk. So all their social life has to take place outside the working environment.

We decided instead to bring people together by replacing the mechanical line with the human work group. In this pattern, employees can act in cooperation, discussing more, deciding among themselves how to organize the work—and, as a result, doing much more. In essence, our approach is based on stimulation rather than restriction. If you view the employees as adults, then you must assume that they will respond to stimulation; if you view them as children, then the assumption is that they need restriction. The intense emphasis on measurement and control in most factories seems to be a manifestation of the latter viewpoint.

## THE VOLVO PHILOSOPHY

Several years ago, when the decentralization and changes in our working patterns were well under way, we decided it was time to disseminate some informal guidelines to help individuals make sure their own decisions were in keeping with our overall objectives.

The issues were discussed widely before the guidelines went into print. Then personal letters were sent to the heads of individual units. They are responsible for knowing the contents and passing them on more widely.

The guidelines are subject to change as conditions change. We published the guidelines informally to keep them from being cast in concrete and then ignored, as so many corporate philosophies are. As they presently stand, I think they have been useful. The following subjects are covered.

**The company's good name** The corporation has a great many human and financial resources invested in its name. The name reflects an image, and that image is the vague but crucial expression of the sum of the company's activities, ambitions, and problems. A name can become a liability, but with forethought it can be made into an asset. The strategic guidelines thus state that the name "Volvo" is one of the corporation's most valuable assets and should be protected and enhanced.

**International operation** We work in more than one hundred nations, with different cultures and value systems. What value system should a company officer use in dealing with this complex world? To avoid dissension, a company should have a clear identity and be prepared to disclose it and live up to it. Thus, the guidelines declare that not only is the mother company Swedish, but the company derives its strength from its Swedish values and identity. This does not, of course, mean that the Swedish values are more precious, more useful, or more sophisticated than any other national value systems. Another important point concerns foreign investment. Because it is a strain to spread your resources thinly across the world, we encourage arrangements where our partners in host countries take a majority interest, so long as they can maintain our rigorous quality control.

**Suppliers** Volvo's supplier system is an integral part of the company's growth and success. The more than 1500 suppliers who provide the company components and technology represent a work force even larger than the 65,000 on Volvo's payroll. To treat our suppliers as we would be treated is not just a "golden rule"—it also makes good business sense. Suppliers are a cornerstone of our total operations, and they share risk with us. It is logical to shop around so we can buy under the best conditions, but these decisions must be founded on very long-term considerations of quality, price, technology, competitiveness, and security of deliveries. We must support our suppliers, try to provide whatever technical assistance they need, and over the long term give them margins that will make them want to develop with Volvo.

**Design** The guidelines state: "The designs of the Volvo group shall be thorough and honest." We want to achieve a technological level second to none, while avoiding the cosmetic frills that so quickly mark the difference between old-fashioned products and timeless products. Continuity is a must, but should not imply conservatism. We want our products to be based on a total concept of transportation, rather than be sub-optimized within any particular sector.

**Information** A prerequisite for our favorable development is to give every employee comprehensive information about the company's total activities, and especially about those affecting the individual's own work. Clearer information can help achieve better accountability, as well as a more balanced distribution of power in the organization.

**Bureaucracy** We can't afford it. Therefore, although adherence to the rules is important, deviations to increase efficiency will be tolerated. Every manager should cooperate, consult with, and try to tap whatever resources exist, but not organize these resources in committees. Finally, the spoken word will always have priority over written memos.

**Production** The ultimate efficiency is achieved when each employee is willing to give his or her best to the corporation. This means we must constantly develop our production methods to give each employee more satisfaction. Every employee is entitled to a dignified workplace and the opportunity to choose variation in his or her work.

The company owes each employee the resources and environment necessary to perform a task, and the opportunity for personal development. Each worker should feel that the company values his or her performance.

**Leadership** Too many people confuse participation and consultation with permissiveness, sloppiness, and a lack of discipline. Others worry about the effects on the organization of increased employee involvement and influence. These factors depend on a new definition of leadership. True leadership builds from an awareness of other people's dignity and their wish to do their best. Good leadership is the ability to cooperate and delegate without losing time or momentum. Such leadership depends on constant and consistent educational programs as well as organization structures that fit the corporation's objectives.

## THE IMPLICATIONS FOR MANAGERS

Leadership is crucial. Participation actually demands better leadership, as well as more self-discipline, from everyone involved. Some foreigners talk about Sweden as if management control, in the traditional sense, may be lost in the new industrial environment. Outsiders also worry that participation by workers will lead to reduced efficiency for corporations. However, examples of inefficiency—and there will be some—are more likely to be due to poor management than to changes in the system. I don't believe that new values and the new laws call for "permissiveness." Instead, managers have to be stronger and more disciplined. It is the weak people in management who have difficulties dealing with employee representatives; until the manager can earn their respect, mutual suspicion blocks an open flow of information. Leaders who have the strength to give and the strength to talk about their mistakes will earn respect. Once the employee representatives trust and respect a manager, real progress is possible. To get respect, a manager must be self-confident enough to give respect to the employees as well. That kind of strength is the focus for selection, training, and development of Volvo managers.

The new work organization methods we have achieved at Volvo clearly demand a different type of leadership at all levels. Foremen, who have been the focus of our production achievements in the past, are now faced with a new situation. The new approach means that they continually run the risk of being squeezed between higher management and workers. Today foremen carry the heaviest responsibility in implementing the new systems. For decades we told them, in essence, that they had two main functions: (1) to supervise the pace of work and keep the line moving; and (2) to give technical advice and assistance wherever necessary. Thus, most of the people who were promoted to foremen's positions had been skilled workers who could solve technical problems. To fulfill the task of seeing that people kept working, they viewed their role as disciplinary, so notices from foremen were full of "thou-shalt-nots."

Suddenly they were asked to develop a rather different set of skills. We wanted them to be "good managers of people." Instead of discipline from the supervisor, the new system emphasizes self-discipline. We redefined the foreman's role rapidly, and this created problems during the change. The problems were exacerbated by the fact that formal training for foremen was traditionally less important than on-the-job training, so they regarded courses warily and with little enthusiasm. Yet, in the new circumstances, foremen needed considerable training to regard themselves as information-gatherers, as aides to the workers, as teachers and consultants rather than bosses. And, in many cases, the attitude change was only partial, stimulated and at the same time hampered because it was forced by pressure from workers and management, rather than from the foremen's own convictions. The situation is better today, but it was a problem initially due to the speed with which we implemented changes, and because we did not consult with foremen enough in planning for the changes.

In a sense, top management can act as an enzyme, a catalyst, to speed up the process of change—but it has not been an easy change, even in Sweden where the social and political values supported the change. In the mid-sixties, there weren't many managers who could support what we are doing now. Managers progressed a long way from the attitudes they held ten years ago, for several reasons. There was force from the bottom, because workers had become so dissatisfied with existing systems, and there was also force from the top. Top management insisted on change. Furthermore, if a middle manager wakes up every morning wondering whether he will have to replace or do without 15 to 20 percent of his work-force, he grows more willing to change. Numbers are important, and the manager has to produce the numbers. The pressure builds up, and this adds considerable force to the pressure from the top. The result has been change much more rapid than the normally conservative culture of a corporation would permit.

Volvo was helped considerably by the fact that we had already redefined the roles of middle management, just as we later redefined the roles of workers. In some ways, control is easier for top management in a decentralized company. In a centralized organization controls can be undermined more easily than in a decentralized company. Instead of saying "cut here," we can now ask specific departments what their own plans are to reduce costs. A decentralized organization with clear boundaries makes improving efficiency much easier. A company experiences less management bickering and more loyalty; employees have a sense that they are necessary. They are better informed, too, with more information about the total operation of their own manageable units, and this tends to give them a clearer perspective on what they are doing and how it fits into the corporate whole.

Decentralizing is most difficult for central management, just as increased participation at the shop-floor level is most difficult for the foremen. And just as the improved shop-floor results are based on better communication, better information, and a sense of membership within the work group, so the improved corporate results in a decentralizing company depend on planning and co-ordination. Therefore, central management resisted the impulse to tighten controls and lay down laws about planning. Instead, some goals and guidelines at Volvo are derived centrally, but how they are achieved—and even what form the planning takes—remains the responsibility of the different units.

The company has no overall "master plan." Rather, the plan, like the organization chart, is decentralized. We expect—and get—a great deal of personal initiative from the people running the various units.

In truth, Volvo is too small to have completely self-supporting units, so our decentralization is not the true decentralization of autonomous units. Torslanda is a major production facility and acts with considerable autonomy, but it is highly dependent on decisions from other sources for such things as its materials—and production in general is dependent on marketing, which is a separate function and a separate organization.

The autonomous Skövde plant builds engines and only engines; it wouldn't be independent if we didn't build cars. Even so, we are able to delegate considerable responsibility to the operating units by making certain assumptions that emphasize their roles as "adults" in the organization.

# THE IMPLICATIONS FOR INDIVIDUALS

Work must be adapted to people, not people to machines. This belief has brought Volvo innovations both in human relations and in technical systems. The new work organization methods permit a greater degree of employee self-reliance. If the individuals exchange jobs, they learn more about each other's work, so the new approach helps them gain a sense of membership in a real work group, the way small groups of craftsmen built cars when Volvo started in 1927. The systems that flourished in the intervening years, on the other hand, prevented communication and group information, and dimished a worker's sense of membership or pride in achievement.

The new work organization, by trying to place people first, also stresses the importance of communication, both upwards and downwards. The individual worker has certain responsibilities as well. Each person recognizes that he or she is dependent on colleagues, both for doing the job and for gaining the knowledge necessary to do other jobs. The new system relies on teamwork, which allows the individual worker to have a greater influence on the overall work situation.

I am sometimes asked whether self-managing work teams are always fair to their members. Do they shun, mistreat, or expel those who are in some respect weaker, or who don't fit easily into the group? These questions concerned us, too, as we were beginning to change. To make sure that inequities do not pass unseen, no team is entirely uncontrolled by management. We have started internal "employment agencies" so workers who have difficulty in adjusting to a particular group, or to the new work situation, will not necessarily be lost.

Education has been a key to much of Sweden's progress in industrial relations. By the end of this decade it is estimated that 90 percent of Sweden's young people will complete high school, and 70 percent will be going on to college. Among these increasingly well-educated people Volvo will have to find its future work force.

The educational changes are not confined to just the young. Even adults are turning to some form of advanced education. At present, about 20 percent of the adult population of Sweden

takes part in continuing education in one form or another. This higher level of education means there will be increasing demands to provide meaningful work in congenial surroundings. The company's response has been to increase its own employee extension courses, in order to make their work more meaningful. These courses not only add perspective to the work employees do, they also help people take on new and more responsible tasks.

Fundamental to Volvo's efforts to make work more meaningful is the concept that an employee's influence on the working situation should be commensurate with his or her capabilities. The experience we have had so far has been encouraging. Employees have shown willingness to take on more responsibility, and their attitudes allow us to take more risks, to make more changes. It is their satisfaction in their work that forms the basis for the continued development of the company. Although Volvo was traditionally developed with resources called capital, technology, and equipment, the company's success in the future will primarily depend on another key resource—its people.

# 2

# Business and its stakeholders

hat is the purpose of business? Very few people today hold to the traditional view that the sole justification for business endeavor is to gain profits for the shareholders. Nor, at the other end of the scale, is there practical justification for the view that the real purpose of business is providing jobs, no matter what the circumstances. There are enough examples of enterprises artificially propped up for this reason, while their prices, costs, and markets go askew, to cast doubt on any such doctrinaire approach. Similarly, it would be hard to back the claim that a business existed simply for the benefit of its suppliers, its customers, its bankers, or the government to which it pays taxes.

There is no single interest group to which business is uniquely accountable—it must be accountable to *all* of the above-mentioned groups, and more. To help handle the conflicts among interest groups and give companies more flexibility to meet the future, I think we need a modern model for business and its stakeholders. This calls for a clear view of the different interests, and some way to think about them harmoniously.

*I believe the purpose of private enterprise is to serve the public.* This single starting point makes it possible to look at the various elements of the public—the people and groups of people —to whom business must answer. Inside a company the task of resolving the public's often conflicting demands is the responsibility of those we call "managers." The following discussion is mainly pertinent to corporations with international activities, but the same sense of accountability and well-identified stakeholders, broadly defined, would help those who must manage smaller enterprises, or public agencies, as well.

## THE GOVERNMENT

In a democratic society, the choice of all the individuals who make up that society is expressed in selecting a government to provide orderly management of the nation's resources and to "manage" the conflicting demands of various groups, just as business management must balance and resolve similar demands at another level. In this context, then, it seems natural that a company should operate according to directives supplied by the government. Where no precise directives exist, private enterprise should work in the spirit expressed by related policies of the government.

This approach can obviously lead to situations in which the interests of the owners of the company conflict with the interests of the public. However, in the long run, I believe it is impossible to defend owner interests that do not align with the country's policies or harmonize with general opinion.

One of the greatest contributions free enterprise makes to society is continuity. More than governments, companies are able to think in the long term; they usually survive both political administrations and dictatorships. Thus serving the public as its interests are expressed by government can scarcely mean that private enterprise must alter its conditions as rapidly as political balances change. The abilities of business to survive and to set long-term objectives are among its most important contributions to the society, and safeguarding these abilities must be regarded as being in the national interest. Thus, a company must be able to follow political changes on the one hand, but not to the detriment of its strength and stability, on the other hand.

The "interests of owners" require reasonably similar corporate structures, and therefore similar corporate behavior, whether the owners are private shareholders or the government itself. Thus, in reality, the difference between nationalized industry and private enterprise is much smaller than most people realize. If a private company exists within a community where government policy dictates public ownership, it should not try to sabotage that policy. This does not mean, though, that the managers and directors should not be free to speak their minds about the consequences of such a change in the ownership structure. My own opinion is that private ownership is the most efficient and appropriate way of running a dynamic business operation.

Governments provide some of the most difficult examples of conflicting demands between interest groups. What happens to a company's activities, for example, when legislators introduce a protectionist trade policy? A company might satisfy the largest number of its stakeholders by circumventing the policy, but this also subverts the interests of the nation and its legally elected government. It is clearcut conflict between the interests of shareholders, employees, customers, and suppliers on the one hand, and the government on the other.

More affluent countries, such as the United States or those in Northern Europe, have developed such mixed objectives that economic growth is certainly no longer given top priority unless the going gets rough. Within such a system, a company has on one hand the role of increasing productivity, yet, on the other hand, must juggle this objective with certain governmental ones —such as social security provisions for strike pay that conflict directly with corporate productivity.

Adolf Berle, in his book *Power,* noted that it is only the combination of ideas and resources that creates power. Resources without ideology result in very little long-term growth. And ideology without resources can be frustrating and sterile. This raises the question of whether corporations, with their considerable resources, should also work towards developing ideologies. With the notable exception of party politics, I believe they should.

In many cases, corporate executives support political parties that espouse conservative ideas, respecting tradition and aiming to preserve private ownership. This is not a corporate ideology, but a personal ideology, loaned to the corporation by owners or managers. However, this adherence to party values can hamper the development of a corporation, because it does not recognize that production and services are there to serve the nation, no matter what party happens to be in power. Furthermore, if a corporation visibly espouses a conservative ideology, it cuts itself off from most labor unions, which have a quite different view. If the corporation avoids the party political trap and can instead develop an ideology of its own that merits support from its many stakeholders, it can dramatically increase its strength. As the situation now stands, on a pragmatic basis alone, party politics, can only damage a company by placing it in a role that can be regarded as hostile to society and polarizing to management/labor relationships.

In patriotic terms, the case against politics is even stronger. I firmly believe that companies should not in any way support political parties. Instead, they should truly stand behind the policy of the elected majority. In this way they can be loyal to the nation.

It is the responsibility of the elected government to spell out the rules under which businesses should operate. And it is the responsibility of corporations to see that their resources are administered within the framework thus defined by government.

By avoiding political entanglements, corporations could do much more than just respond defensively to policy changes proposed by politicians. People have a need for clear exploration of social issues. With a few notable exceptions, corporations have not done enough to initiate debate on a wider front, to raise questions beyond corporate boundaries. If a useful dialogue is to be established between government and corporations, it will need to be based not on party politics but on good sense and credibility on both sides, and on a shared awareness of the real social issues.

If one accepts that the purpose of business is to help achieve and maintain the public good, then it is logical that a further objective must be to administer the resources with which the company is entrusted and use them to create economic growth, taking into consideration all the interest groups involved with the company. This objective carries with it the demand to provide meaningful employment.

As it becomes clear that rapid exploitation of resources has strained not only the environment but also the people who contribute so much to the growth, the objective of growth has received more critical attention. During the fuel crisis in 1973, for example, many leading statesmen in the industrial countries gave somber warnings that our growth epoch was coming to an end, that the consumer society lay behind us, that new values were making themselves felt. Countries were obliged to learn to handle stagnating economies. Not many weeks passed before the governments of the industrial countries began gearing up their industrial activities to full speed to create the funds necessary to pay for the ever-more-expensive energy. The growth society had been given something to think about, but the real course of events did not waver much. As long as the aim of society is economic growth, then private enterprise will be the driving force behind growth.

If we accept that the objective of an enterprise from the viewpoint of society is to administer the resources entrusted to it and to create economic growth and thus generate employment, then within this framework a company must draw up its own objectives, based on the demands of all its different interest groups.

## THE CONTRIBUTION OF CAPITAL

The influence of financiers, bankers, and industrialists, especially if measured in terms of their public images, would seem to have declined in recent years. Yet these people remain an important factor in the power structure of most countries. Both their power and their image problems are reflected in such terms as, "the military/industrial complex" in the United States, " the establishment" in Britain, or "le patronat" in France.

The public generally uses the term "business establishment" to refer to a collection of directors, entrepreneurs, financiers, and shareholders, a group which, by implication, is assumed to pursue activities that are not wholly beneficial to society. Yet with all the criticism, one fact is inescapable. No matter what the social system, every country needs a workable structure of business enterprise. Our capacity to produce goods and services is, after all, the cornerstone of our living standards and a vital nerve in every community.

Volvo, for example, has about 65,000 employees, which makes it Sweden's largest private-sector employer. Volvo is responsible for almost 3 percent of the country's total employment and 7 to 9 percent of its export. Our 1500 suppliers employ more people than we do. Volvo has almost twice as many shareholders as employees (about 120,000 shareholders) and at least 70 percent of them hold fewer than 500 shares each. The welfare of all the shareholders and employees depends entirely on our ability to maintain a profitable and competitive business. Thus we have set ourselves the long-term objective of 15 percent return on capital employed.

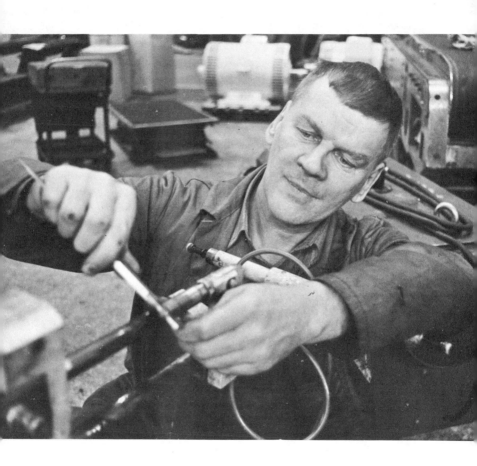

The company's owners, the shareholders, initiate its activities and appoint its board (and thus its management). Within the framework of the law, they invest the capital which promotes continued stability in the rest of our social system. According to the legislation governing enterprise (which itself expresses the public priorities), these owners have a self-evident right to demand decent returns on the capital they venture in the enterprise. It is by satisfying their desire for profit or by convincing them that they will do even better by reinvesting some profit towards growth that a company has the capital to grow and thus satisfy the demands of other interest groups. Without their capital the company could not subsist, because growth requires investment. Thus, a primary objective of each company is to achieve a competitive return on the capital its shareholders have invested. Until we find a better yardstick for measuring how efficiently resources have been utilized, return on investment will continue to be the criterion on which the financial community bases its investment decisions.

A company's management must not be regarded as an exclusive representative of the owners, though. This implies that in any given situation the management would place the owners' interests above those of other interest groups. The operation would thus become asocial, and would not serve the broad interests which are the company's full mandate. Instead, the task of management is to maintain the productivity and efficiency that permit all these interests to be met with the least possible conflict. The owners must, in all fairness, be able to demand that they not be forced out or squeezed so hard that their fundamental requirement of a good yield cannot be satisfied. Such a guarantee is also in society's best interests, so long as we recognize that private ownership and competition are basic conditions for a dynamic economy.

The shareholders thus pay for the growth of the company. Through the capital they provide and the risk they are willing to accept, they also create the platform for the company's borrowing from other credit sources. In today's world, the profile of these owners makes up the business card of a company in the international capital market. By submitting their company to intensive, ongoing scrutiny, through its public listing on stock exchanges and in the business news, they help keep the management open to viewpoints and requirements not only from the owners themselves but from the business environment in general.

A decline in the importance of shareholders became noticeable during the growth period of the sixties. However, they are more highly esteemed today as problems in the business environment reveal more clearly the true importance of their capital. Today corporations recognize the value of their continued support.

Shareholders contribute the company's own capital, and risk losing it entirely if the venture is not successful. In exchange for this risk, they have the opportunity for very good yields. The institutions that grant credit take smaller risks, because their loans must be repaid before shareholders can receive their money from a venture. On the other hand, the credit-granting institutions charge less for their capital than the dividends shareholders expect from a prospering company. The lenders demand that a company must be able to repay both the interest and the capital they lend; thus, they require a sound and balanced growth picture, bolstered by pronounced profitability. In this respect, the lenders and the shareholders have similar interests, with a bias towards stability from the lending institutions and towards risk from the shareholders. In addition, the lending institutions in modern society are also answerable to the public in ways that individual or institutional shareholders are not. Thus the lenders, at least indirectly, demand that a company be a good corporate citizen with faultless ethics. When these demands are met and good relationships are firmly established, they can support a company with loans for its growth and also help when help is most needed, notably when the company encounters difficulties or experiences a downturn in profitability.

## THE CUSTOMERS

Just as vital to the success of an enterprise as its shareholders and lenders are the customers to whom its products must appeal. It is their demands that determine the directions for research and development, product design, marketing, and service. Obviously, if the demands of the customers are not met, none of the demands of any other involved interest groups can be met in the long run.

In the case of a company like Volvo, with a range of mass-market products, many employees, a broad share ownership, and many subcontractors, customer demands are often synonymous with those of other interest groups, simply because members of those other groups are also customers.

Implicit in the customer/company relationship are certain responsibilities the company should assume:

- *It should view its customers as adults, capable of making informed choices, rather than as children or ciphers who can be exploited or manipulated.*

- *The company owes its customers purpose-built products and properly adapted services.*

- *The company should be able to guarantee the durability of the product, and to give the customer fair security against any mistakes that might have been made.*

The best medium I can see to assure both companies and customers that these responsibilities are met is the free-enterprise system—i.e., competition. So long as the customer has other sources from which to choose, the company has a strong motivation to retain customer loyalty and gain new customers by improving the quality of its products and services. It is from this customer viewpoint that I particularly reject the idea of state ownership, which by its nature strangles competition and restricts the selection of products available. Even if it were theoretically feasible to imagine a state-controlled business that maintains ambitious targets for its customers, it could never embody the alertness that stiff competition serves to generate in private companies.

## THE SUPPLIERS

Practically no company is so self-sufficient that it controls all the basic products and components needed for its own production. Thus, a company's subcontractors or suppliers are in most cases another essential interest group.

The company demands that its subcontractors produce parts and components or provide services of a specified quality, at the right time and for the right price. The company can also demand a certain flexibility from its suppliers—e.g., that they relax demands for payment during a crucial period, with the chance for expanded business when the company is back on its feet.

For their part, the subcontractors have the right to demand certain assurances from the company:

- *They deserve the best information possible concerning both the company's long-term plans and its payment capabilities.*

- *They have the right to expect that the company will share its technical knowledge; the firm that assembles and markets the final product has a better overview of the technical development and market demands than the company which makes only components.*

- *They should be able to demand that a major customer be a strong company with good prospects for survival.*

A subcontractor needs a certain amount of security, too. This comes not only through advance warning when a difficult period is approaching, but also through arranging activities so the subcontracting firm is not entirely dependent on a single customer. There is considerable merit in the corporate policies that state that no single subcontractor shall have more than a specified portion of his business from the company—a case of the corporation limiting the business it puts out to subcontractors, in the subcontractors' own best interests. These limits usually range from 10 to 33 percent, depending on the type of business.

# THE INTERNATIONAL BUSINESS COMMUNITY

It goes without saying that the countries in which a company operates should also be listed as interest groups. However, these are difficult to define because they usually incorporate governments, customers, shareholders, lenders, and almost always employees. This underscores the fact that a company actually works in an international context.

While government clearly belongs to the nation, business is increasingly international. Though few companies are actually "multinational," with owners, managers, and employees in many different countries, the term is becoming derogatory, and it is also misapplied to companies that simply export or import goods or services, or maintain a few foreign satellites.

We are brought up to be loyal citizens of our countries. Like animals suspicious of different species, people are naturally suspicious of strangers, and someone from another nation is strange in many ways. Thus the government department is our own, but the company that just reported a profit is suspect. Whose jobs have they exported to achieve that profit? What employees have they overworked and underpaid? How much did they overcharge us for their products? The questions may seem absurd to the businessperson, but because people are asking them, they have importance. Unless such questions are answered clearly, openly, honestly, and often, people will continue to ask them. Business needs to do a better job of explaining itself generally; when business is international, the need is increased.

There are fair answers to the international suspicions. The truth is that economic activity is far more international than political activity. In today's closely interlinked world, it is the politicians who ought to be apologizing for their incapacity for international understanding, rather than the business people apologizing for the stigma of the multinational.

International business brings a number of benefits to a country. In answering to the "job exporting" accusations, for example, an IBM executive in 1973 told a United States Senate subcommittee that one out of every five jobs in his company's

United States plants existed because of foreign business. A United States Tariff Commission study the same year reported that the American companies most active in production abroad were also the heaviest contributors to United States exports.

The classic argument is that international companies can put capital to work in countries where capital resources are scarce. This is true and useful. But I see other equally important benefits. A company that is active internationally is also able to transfer technology from one country to another. This is not a "zero sum game" in which the transferring country loses that which is transferred. Instead, both countries may benefit. The headquarters company and country have access to new ideas from people knowledgeable about a different culture. The receiving country gains new skills. The people who move gain training and a broader view of their special areas. The world gains an increased pool of international understanding and a slight reduction in the natural suspicion of the unknown.

These are arguments that business could illustrate with vivid examples, yet we rarely see them. It is easy to understand the corporate fear of exposure, the risk of being misunderstood. But it is better to be misunderstood for being open than for hiding things that need not be hidden. Where is the TV interview with the engineer who just came back from a three-year assignment in Nairobi? Instead we hear from the politician coming back from a two-day official visit, or the travel writer who never left the tour group.

To be open, we have to be candid about our black spots as well as our achievements. There are abuses in business, particularly in international business, because multinational companies are in a position to play countries off against each other. A few companies have enormous power. They do business on a global scale. They can, in isolated cases, endanger even governments by withholding necessary goods or involving themselves in local politics. At present, the largest international corporations are virtually unaccountable to anyone.

A few companies tend to congregate in places where resistance is lowest, where taxes are lowest, where wages are lowest, and where social pressure is lightest. This happens because a firm may occasionally lose sight of objectives other than profit that would assure them longer, healthier organizational life. Their years of foreign experience have helped most firms build up rules of behavior and ethics for long-term success. These have developed into an international community of values. For the company to prosper, the host country should prosper. The company's contributions include fair taxes, fair wages, and upsetting the currency or import/export balances as little as possible. Many companies do operate this way, but when one maverick monolith misbehaves, other firms tend to ignore the abuse, silently helping to sweep it under the carpet. An organization's defenses are always up when the organization is under attack, and, in this sense, the international business community often behaves like a superorganization.

Instead, the few abuses could be used more positively, to speed up a set of common agreements, spelled out with governments and the public, on such aspects of business as personnel policy, taxation, location of facilities, accounting, accountability, wages, social benefits, research and development, conditions of production, transfers of foreign currency, or repatriation of profits.

Until such agreements are achieved and generally understood, some companies and some countries will continue to exploit the anomalies for their own short-term benefit. Most of the blame for such behavior will fall not on the countries but on the companies, and by association on all companies.

Most international firms are already behaving according to such ethics. The main change, which would be uncomfortable for many, would be a requirement for increased and more standardized accountability, not just to shareholders but also to the public. In this time of increased concern about personal privacy, one luxury that business may have to forgo is corporate privacy.

Conflicting demands among interest groups are particularly prevalent in international business, partly because of the natural suspicion people harbor of foreigners. An internationally active company often finds itself involved in a dilemma. To succeed it must abide by the values of the host country. At the same time, it has to take into account the policies and opinions in its home country. Unless they represent particularly large markets, it is difficult for foreign governments to "bully" a large company, but it is much too easy for the home country, which usually includes the company's owners and is limited to national perspectives, to treat unjustly the company's interest groups in other countries.

This balance between interest groups is a classic example of the increasingly complex conflicts managers are expected to resolve. What happens, for example, when a company has been established for a long time in a foreign country, and then a change in government there causes adverse reaction in the home country? Somehow the company must demonstrate that it has listened to the political signals from home. Yet, at the same time, it is seldom prudent to make rapid and abrupt changes in its activities abroad. In the case of a new dictatorship, for example, the lives of employees might be at stake if a company were to withdraw immediately. There are also customers in the host country who have trusted the company to honor its warranty and service obligations. Management has to consider the long-term interests of all groups, and take into consideration the fact that a change of government can take place yet again. Generally speaking, a company should not close down its activities as a result of a government change for the worse. Instead, it should probably continue its basic activities, but refrain from making new investments or acting in any manner that could be regarded as encouraging to, or supportive of, the administration of the host country.

This is sometimes a question even before investment is made. Should a company refrain from setting up shop in a developing country because people in the home country disapprove of the regime of the developing country? There is no easy answer here, but one thing seems clear: a company should not conduct a foreign policy of its own, any more than it should be involved in party politics at home. An international company is not the proper institution to formulate social objectives for others. This is a political, not a corporate question. The consequences of a company behaving as a political power factor would be far worse than the consequences of a company acting on principles of pure economics.

On a purely economic basis, there is one useful rule of thumb. For investment to succeed, it should harmonize with the needs and values of the host country and should be aimed at having a positive effect there. The moral aspect a company director has to take into consideration cannot be determined solely by the values of one's native country. Instead, the company must follow a course of action which is an acceptable compromise between its own efficiency and economic criteria, the values of the host country, and the attitudes in the native country. If these cannot be resolved, the investment is unlikely to succeed. The most important guideline would be for a company never to contribute deliberately to activities that are contrary to fundamental principles concerning human rights.

In the United States, Sweden, and other advanced industrial nations, the economic recessions in the seventies aroused some concern that manufacturing investment outside one's own borders meant somehow exporting job opportunities. In some instances, a company can choose between an investment that will increase employment at home or a corresponding investment elsewhere, but in most cases the situation is more complicated and various factors, such as development and production costs, cooperative agreements with foreign firms, or putting final assembly as near the market as possible, more or less compel the company to invest abroad. Many sound companies in small home markets could not sustain their activities by domestic activities alone. To survive they must expand into foreign markets to attain sufficient volumes. It is no coincidence that several of the top companies on *Fortune*'s international 500 list are jointly owned by the Dutch and the British.

Most countries do not live in isolation. They depend on their exports and their business activities abroad. Even in the United States, where international business represents only 5 percent of the GNP, there is a growing dependence on oil imports that have to be balanced by exports. To improve their living standards, countries have to recognize and accept the economic realities and the conditions under which international commerce operates. No country is alone in wanting the greatest number of job opportunities and the greatest volume of exports. So companies have to strike a balance—they must satisfy the demands of their native countries without becoming unwelcome guests in their host countries. Only when this principle is accepted can a stable foundation be laid for future business.

## THE EMPLOYEES

The conditions society sets for free enterprise today include an increasing emphasis on the special interests of employees. One of the major social requirements is that work shall be available to everyone. Continuing cutbacks in working hours and increases in holidays imply that everyone shall work less. The increasing productivity is to be divided so that everyone gainfully employed will get better pay. The endeavor to achieve democracy in working life carries demands that all citizens shall have greater influence not only on their own tasks at work but also on the total operations of their companies. A new demand now coming to the fore in some European countries is that employees not only should have substantial sick pay, but that they should also be able to decide for themselves whether to stay at home. On top of these demands comes, finally, the fact that most governments try to distribute employment geographically—a demand not only for work for everyone but also for work for everyone everywhere.

Job mobility, which has been taken for granted in America for some time, is a relatively new phenomenon in Europe. America experienced immigration for many years, while Europe since the late 1800s has largely been accustomed to emigration. A number of European countries in recent years have encouraged immigration from southern latitudes, to provide additional labor for their most productive economic sectors. Job mobility within countries first became apparent during the 1950s and 1960s. Farming was nationalized during and after World War II, causing a considerable proportion of the traditional farming population to move from the land to industry. Thus, expanding companies had access to the labor they needed and people moved within their countries from less productive to more productive sectors of the economy. This helped create the basis for the vigorous industrial growth in the fifties and sixties.

At the same time, this high mobility created social problems. People did not always adapt quickly to entirely new environments and activities. Those regions which grew explosively and absorbed immigrants were not always able to manage their growth harmoniously. The rate of job mobility increased so much it finally resulted in high personnel turnover within industry and subsequent problems of maintaining quality standards in production.

The major motivating factors for growth during the fifties and sixties have changed drastically, and in the seventies we encounter an entirely new set of demands for the benefit of employees:

- *the demand for meaningful work,*
- *the demand for job security,*
- *the demand for competitive pay,*
- *the demand to be able to stay in the same place,*
- *the demand to continue in the same type of work,*
- *the right to determine sick leave for oneself, and*
- *the demands for influence and co-participation.*

Let us look at each of these demands a little more closely. Occasionally they conflict with each other, or with demands of other interest groups, but resolving the conflicts is management's challenge today.

Meaningful work requires new methods of manufacture, of organization, and of management. Understandably, these requirements also increase the need for good communication of essential information both about the work the individual does as well as his or her relationship to others in the group and the company. As I see it, the demand for meaningful employment is not something that will act as a brake on development. Handled correctly, it can stimulate growth in a society where people are increasingly well-educated and competent. It is important that their competence be used, because it is a vital resource.

The first and best way to satisfy the demand for job security is through growth that is coupled to real consideration for the individual. Even though this kind of consideration might be viewed as a form of restriction on corporate freedom of action, I believe that the demand for job security, like the demand for meaningful work, will be a stimulant to industry rather than a drawback in the long term. However, security should not be used as the excuse to fight tooth and nail to save an operation which must, in reality, be phased out because it is unable to grow or to contribute profits. That would ultimately endanger other jobs as well. Instead, security can be assured by means of a "social cushion" that is corporate-wide, and by achieving confidence among employees that private enterprise as a whole and their own company in particular can compete and thereby flourish.

The demand for pay that matches or exceeds pay offered by similar companies is best satisfied through sound growth. Here again, demanding competitive payment from an operation which cannot support it gives only false security. That operation will, of course, ultimately fail, to the benefit of a sounder or more substantial competitor. From the worker's point of view, the demand for fair pay is justified. More broadly, labor demands for a share in the company's success amount to a type of profit sharing. Some people feel these demands should be satisfied within the company's remuneration capabilities. However, as unions grow and demand a levelling out of pay scales between different activities, the chances for differentiation dwindle, and with them the opportunity more profitable companies have to pay higher wages than their less profitable competitors. Thus wages will become less an expression of a company's total payment capabilities. A profit-sharing system should, in principle, be able to stimulate business and its development.

The demand for stable work, at the same place, can clearly impede structural development, and is thus opposed to the demands and aims of other interest groups on occasion. Our society should be arranged so that reasonable respect is paid to a person's desire to live in a certain place, without resorting to the ultimate price of artificially sustaining unprofitable enterprises.

The increasing demand for assurance that a person can go on doing exactly the same type of work is probably a reaction to some excesses of job redesign, and will hopefully taper off a little as more human organizations evolve and people have more control over their own working lives. If this requirement has to be put into the balance against the demands for individual development, then I believe the latter deserve more weight. If a specific job is eliminated, it is important that the individual have every assistance in finding a new job that is equally satisfying. But a stubborn adherence to outdated types of work would gradually bring any enterprise to a premature geriatric state.

The right to be sick and the right to take time off can never be opposed from a humanitarian viewpoint. Quite simply, this has to do with the freedom of the individual. Any restriction placed on the right to take time off would have to be decided through the processes of government: How much does this right cost, and how much can we afford to pay for it?

The demands for influence and co-participation can inhibit industry if they reach the stage where the organization cannot function smoothly or dynamically because continual halts occur in the decision process to satisfy employee demands for information and influence. On the other hand, if it is possible to create room for such influence within the framework of a flexible organization, without causing injury to the decision process, such demands can stimulate an organization, particularly when the individual feels personal development opportunities are increasing and when he or she gets a better understanding of the needs of the enterprise. This can lead to greater harmony among the demands made by the labor force and the desires of other interest groups.

Labor's demands for a share in a company's financial success exemplify the many factors that must be taken into account in meeting the needs of all the interest groups concerned with a company. They also underscore the shift of power that has taken place along with the increasing interrelationship and interdependence of interest groups.

There is a definite trend in Europe today toward the so-called "solidarity" wage policy. This means that people working in a branch or firm with poor profits should have wages not significantly lower than those who work in companies with good profits. This demand is based on the ideological view that every employee is entitled to a good basic income, and that it is not the employee's fault if a certain enterprise is losing money. Nor does the employee have any chance to turn losses into profits. That is viewed as a management responsibility.

But solidarity brings problems. The demand for equal pay erodes the difference between the highest paid and the lowest paid workers. If the minimum wage is too high there is a risk of killing off companies that would otherwise be able to survive and do a good job, offering employment, dividends, and so on. If the maximum wage is too low, the employees' share of the pie has to drop, to the benefit of other interest groups. The more we progress towards solidarity, the more prominent grows the employee-based demand to get an extra share of the financial returns from the companies that are doing best.

These demands have led to profit-sharing discussions in a number of European countries as well as developing countries. Profits can, of course, be shared in the form of cash, which would provide some motivation to employees, although it would remove capital from business that could use it for further investment. A new system under discussion would allow profits to be distributed in the form of shares in a fund, which holds corporate shares issued for the benefit of the employees. Various objections have been raised to such a system. Employees point out that only those employed by a company would share in that company's profits. Shareholders express some concern that their holdings may be diluted to the benefit of employees. A system of profit sharing through share issues could lead to a complete takeover of a company by labor organizations, unless a maximum limit were set on their shareholdings. Furthermore, if such a scheme were launched for employees in the private sector, one could expect an immediate demand for higher wages in the profit-free public sector, leading to higher wages in the labor market as a whole. This could continue to distort the entire wage structure and create a drain on free enterprise to the extent that it would no longer be profitable, or that the interests of other groups could not be satisfied.

To its credit, a profit-sharing system expressed in the form of corporate investment would probably encourage employees to show caution in their wage demands, thus leaving more money in the company for continued investment. Furthermore, such a system could lead to an increase in corporate efficiency because the employees, being shareholders, would participate in cost-cutting exercises and perhaps invest in new share issues beyond their normal profit-sharing holdings.

If the public sector is to take part in this kind of development, they will have to have a similar kind of savings opportunity perhaps by setting aside some part of their salaries just as employees in various fields sometimes form savings or investment clubs.

I personally am no advocate of collective funds, which might mean that an entirely new interest group led by a few labor unions would have a dominant influence on all private enterprise. I prefer a vigorous individual participation, permitting people to invest in corporate shares to the extent they wish—an approach that is already common in America today. This would first of all stimulate individuals to take part in the development of business and economic life. Second, it would spread out share ownership to the benefit of both business and society.

## WALKING THE TIGHTROPE

Private enterprise today is tremendously complex, not only in the work it does but also in its organization and structure. The number of interest groups has grown, and the desires and needs of each one are many-faceted. While companies have tended to increase their scope, doing business on a larger scale to achieve lower unit costs and sustain profits, the restrictions on their freedom of movement have increased.

The problems for management, the delegate of all these interest groups, have multiplied accordingly. Integrating these complex demands has sometimes made the whole structure so inflexible that a new model of the stakeholders is called for. If management does something that affects one group today, it finds a chain reaction setting in among the others. Business has taken on a new form, and solving today's problems of conflicting demands requires new approaches. For management it is a major challenge to resolve everyday conflicts without getting caught in the middle of the battlefield in some classic class struggle.

The shift of power toward labor has revealed how fragile is our concept of "management," and how ill-equipped traditional models leave us when we must deal with the complex conflicts that arise among independent groups that have a stake in a corporation.

Although the problems are immense, I am convinced that a new model, calling for keen hearing and sensitivity to demands from the surrounding world, is necessary for future survival.

When industry was in its infancy, the profit demands of shareholders were the dominant force and, as a result, company managers were fairly accurate in their view that they were in the employ of the shareholders. But as the decades have passed, other interest groups have gradually emerged. For a long time these co-existed in a kind of natural, matter-of-fact harmony. Management could still view its primary task as satisfying owner demands, and in so doing automatically satisfied the demands of other interest groups.

Today, though, the demands collide more often. Each group must learn to view the company as a separate entity, rather than an extension of itself (or an extension of the owners).

Accordingly, the manager's boss is no longer "the owners" but instead "the public." The manager can no longer be an exclusive instrument for any single interest group. His or her job is to create growth and oversee a well-balanced division of the yield between the various often-competing interest groups.

This raises the question of who hires the manager. In most countries, corporate law stipulates that the board of directors is entitled to appoint operating management. This is fair, because the board in most cases has to take responsibility for the results the managers achieve. Traditionally the board was appointed by shareholders, but more often these days laws have been amended (or companies have taken initiative) to include employee representatives on the board, and in some countries and industries the government itself has representation on a company board. The and steel industries of West Germany have equal shareholder/

employee representation on their boards, and banks and insurance companies in Sweden have government representatives.

I personally believe that shareholdings are so broadly based today that shareholder boards are fairly representative. But no matter who appoints the board, if a company is to have a mandate to continue its activities, it must behave as a good citizen. It is the responsibility of the company's board and management to ensure that the enterprise keeps the public interest, in its broadest sense, as the primary objective of the business. Only in this way can all the demands of all the interest groups be satisfied in the long term.

# 3

# Kalmar:
# the catalyst

T o understand what happened at Kalmar and later at our other factories, it helps to go back in the company's hitory for a moment. Volvo started in Gothenburg in 1927. Like other companies in the youthful auto industry, our production was based on groups of skilled craftsmen working on a single car until they finished it and drove it out the door. The primitive "production planning system" judged productivity by simply counting the products as they came out. The products themselves were fairly simple, too—cars, trucks, and buses.

Foremen were the key to production from the start. They were skilled men who had been promoted because of their technical skills. They were expected to develop authoritarian and paternalistic attitudes towards their workers, and they quickly lived up to expectations. In the traditional sense, morale was high among the skilled, stable work force. It was taken for granted that you did a good day's work for your money. Management as we know it today was not necessary. You managed the system, not the people.

## THE CHANGES

By the end of the forties, growth had somewhat eroded the group approach and technology was growing more important. In 1953 Volvo introduced the MTM planning system imported from the United States. This became the company's key planning instrument and still remains so. During the fifties, Volvo became heavily mechanized. Group work continued to decline in the face of increasing specialization and shorter work cycles—the classic pattern in the auto industry worldwide. The human side of planning never merited much emphasis because economic growth was the company's prime objectives, as it was in most other countries at the time. If you got growth, you could pay people more, and everyone knew they would then work hard and be happy—or so it seemed.

Volvo's growth continued throughout the sixties, but by 1968 it was becoming obvious that there were problems ahead. The big Gothenburg plant, Torslanda, encountered wildcat strikes, and sometimes these were echoed in smaller factories too. By 1969 the employee turnover rate reached 52 percent and absenteeism was increasing. Management began to pay more attention to social, sociological, and psychological factors.

As the seventies arrived, each plant began to develop more human and humane working patterns, but each in its own way, due to differences in plant size, products, production technology, management style, organization, and factors in the local environment, such as availability of labor, use of foreign workers, urbanization, and cultural traditions. In general, though, most plants at this stage gained broad experience in joint consultation with employees, as well as work organization and some development in the social and technical aspects of work.

Although the company's first employee representative body or "works council" was established as early as 1946, this kind of group grew more important in the seventies. Volvo began to develop a works council structure paralleling the corporate organization. Works councils are today the backbone of the company's relationship with its employees. The network of employee-elected groups now stretches from the board level Corporate Works Council to smaller councils for specific departments inside a single plant. Joint project groups or consultation groups are the links between the corporate and works-council structures at every level.

The goal of all this consultation has been to improve the information available to the employees and the company alike. We believe this goal is being met and extended every day.

Job rotation was introduced in Volvo in 1964, but, for several years, was conducted on a very small scale. In 1969, after the first wildcat strikes, job rotation became more important. A joint employee/management consultation group set up to prevent further trouble recommended that job rotation be emphasized.

In general, the situation seemed volatile when I arrived at Volvo in 1971 from a completely different industry. Export sales were going up rapidly, but most of Volvo's production was still in Sweden. We could see the end of the era of the production engineer with technologists designing the plants, the cars, even the tasks of individual workers. There were also changes in the complexion of the work force that created concern and uncertainty. The difficulty in hiring Swedish workers meant that most companies had to look for foreign workers.

## THE RISKS

The Kalmar project was initially a normal planning exercise for a new but traditional plant. The planners were working with more emphasis than ever before on good working conditions, but no one had considered major departures from the norm in the new assembly plant we needed. Kalmar had been chosen as the site because it was a pleasant place with easy communication and transportation links to our other facilities, and skilled workers were readily available. Recent cutbacks in another factory there meant we would be a welcome addition to the depressed local economy.

The planners were already working on the premise that humanity and efficiency could be combined to some extent. Their first report emphasized group work, expanded the tasks that each worker should do, and gave work groups some opportunity to vary the speed of their work. It focused on good physical conditions, ample space, and noise control, but still included the conveyor line.

A new factory, though, presents a unique opportunity to try out entirely new solutions. Here, starting from scratch, we might be able to achieve changes that would be very difficult to implement within an existing plant. Therefore, quite late in the planning cycle, I interrupted the project and set out new objectives. A new and non-traditional type of task force was pulled together and given the almost impossible task of designing a landmark factory as an alternative to the traditional plan.

The ideal goal for the new plan was to make it possible for an employee to see a blue Volvo driving down the street and say to himself: "I made that car." The original objectives for the alternative planning group assumed that they could develop a new materials handling system and a new product transportation system. The design must give individuals as much control as possible over their own working lives.

The guidelines memo concluded that instead of a conveyor belt moving through a warehouse, Kalmar should be based on stationary work, with the materials brought to the work station. Each group work area should accommodate about fifteen workers. Tasks could be varied within a group, and each group would take more responsibility for the quality of its own work. We also felt that the strong reputation of our products in the marketplace should be reflected more directly back to the workers. This meant that no one's view should be limited to nuts and bolts, or single components.

There was nothing earth-shattering about these ideas and suggestions. Many chief executives have had similar ideas. The difficulty lay in getting a large and successful company to make changes that were not necessary in everyone's eyes. The traditional factory plan had been completely acceptable to managers and workers; it entailed few risks.

Even exploring the nontraditional approach involved a number of risks, especially in 1971 when nontraditional approaches were rare. The very existence of the alternative task force first of all disparaged the work of the traditional team, and thus of traditional production in general. This is not a welcome factor in the culture of any auto company. If we went ahead with the nontraditional approach, we incurred the danger of involving reluctant members of the company in risky, untried production methods about which they felt neither confidence nor enthusiasm.

If this project went wrong, it might hamper less radical changes in other parts of the company, and actually hold back our progress in making work more human. This was cause for serious concern. Almost all of what Volvo stood for in production was reliability, the tried-and-true approach—an image that carried over into our products. Thus, it was not surprising that most production people initially were aligned against the venture, although the Kalmar project, originated at the top, gained support of a sort.

The risk was international in its scope. We carried the responsibility of proving—or failing to prove—to the industrialized world that there were realistic alternatives to the traditional assembly line.

Looking at the risks, top management decided to go ahead with the nontraditional Kalmar project anyway. In my first meeting with the new task force, I had looked at the long, straight, inflexible lines of traditional assembly and sketched the layout in a number of circles, with production around the edges and materials in the middle. Such a solution should give back the sense of involvement. The task force had only two weeks, starting from this vague sketch and a lot of seemingly conflicting and idealistic guidelines, to develop an integrated concept based on the new principles.

The project leader knew there was risk involved for him, too, and there was pressure to meet the original schedule.

The project leader was particularly good at picking people. He pulled together a team of eleven, including managers, production engineers, the architect, and a foremen's representative. Working at full speed, they produced a plan that solved the problems, at least on paper, and included an entirely new way to handle the flow of materials within the factory.

Their draft plan was accepted and the board voted to go ahead with the project, even though it entailed an initial investment that was about 10 percent higher than the plan for the traditional factory at Kalmar. We had scheduled the first car to come off the line twenty months later. To give the new approach credibility within the company, that schedule had to be met. The project group believed its concept could be implemented without danger to the schedule, including development of the new materials handling system. Production costs—man-hours per car—would be about the same under either plan.

## THE KALMAR CARRIER

The heart of the Kalmar technology—and of technical advances in other factories since then—is the carrier on which assembly takes place. Instead of the fixed-track conveyor belt, the project group sketched out and Volvo engineers created a new carrier, which is now a patented product in its own right.

An empty carrier looks like what it is, a low platform, quite simple, slightly larger than a car. The carrier has a corrugated

*Workers can control the carriers for assembly.*

*High-level carriers are used for engine, gearbox, axles, and exhaust assembly.*

metal surface for safe footing and a narrow waist so workers can stand on the floor if they choose.

The carrier itself gives flexibility. It is a self-propelling vehicle that runs around the factory according to a pattern of conductive tape in grooves on the floor. The tape can be taken up or put down easily. Manual controls can override the taped route as well. The number of carriers in use at any time can be varied, permitting further production flexibility.

The carrier for Kalmar comes into two versions. The low-level carrier is used for most of the assembly. The high-level carrier is especially designed for the lower parts of the car, so group members can work in normal postures. The low-level carrier has a device that turns a car 90 degrees, so workers can reach its bottom easily and comfortably.

Driven by electric motors, the carriers can be operated in a number of ways. A cluster of small computers at the factory sends impulses through the tapes in the floor for normal progression through the plant. Each team can override the computer-generated signals from a control panel in its own area. And because each carrier has rechargeable batteries in addition to the tape power source, a worker can stop a carrier or change its path —off the tape, if need be—by using a small nine-button controller on the carrier itself. The system is so flexible that there is nothing to stop a work group from getting on a carrier and going down the road for a picnic. The carriers normally move between stations in the factory at about one mile an hour. Each one has a spring-loaded bumper, so if it runs into an obstruction it bumps gently and stops, moving on a few seconds after the obstruction is removed. Visitors often test this by standing in the path of a carrier. As yet, we have no claims for bruises.

The first glimpse of the factory with its carriers is slightly strange for a visitor. Here and there individual carriers start or stop, seemingly of their own accord, or make regal moves through complicated turns. Their usefulness becomes clear as you follow the normal production cycle.

Painted bodies arrive every night on freight cars from Gothenburg while other parts come directly from Volvo suppliers and other plants. After cleaning and inspection, each body is mounted on a low-level carrier. The computers automatically send the carrier up an elevator to the top floor and log the body in, printing out the specifications for that particular car.

Special sensors report the position of each carrier as it moves throughout assembly. The computers also send special assembly instructions for each car to terminals in each assembly group area before the car gets there, so by the time a carrier arrives at a station, its team has already had time to make up the subsidiary components it puts onto the body there. Using the computers to keep track of the carriers means that the carriers do not have to follow predetermined paths in a certain frozen order through the stations. If someone notices a scratch in the paint, the carrier can be turned back to the painting station immediately, and can later return to the appropriate production process.

The door shells are taken off and assembled separately while the body goes through the ten team areas upstairs, getting its controls, safety features, glass, headlining, and electrical systems. Then the completed doors are installed.

In the meantime, a high-level carrier on the ground floor has been passing through other team areas for assembly of the engine, the gearbox, the axles, and the exhaust system. Then the high-level carrier goes up in the elevator, the completed body is mounted on special supports just above the chassis, and the entire rig returns to the ground floor where a special team mates the chassis to the body—the most dramatic part of the assembly process. After this, the body supports are folded out of the way, making the nearly finished car accessible so the next team can fit the brakes and wheels, fill up the brake fluid, and test the brake pressure. Then the car is moved onto a low carrier for its final accessories, final tests and adjustment, and delivery approval. The assembly process finishes with underbody sealing and a final coat of protective wax.

## THE BUILDING

The design of the plant, like the design of the carrier, is focused entirely on the working groups. Instead of the original circle I sketched, the project group and architect Ove Swärt came up

*Hexagonal plant design gives each group its own wall and windows.*

*Cutaway shows how assembly flows around the outer perimeter of the Kalmar plant, while internal areas are used for materials.*

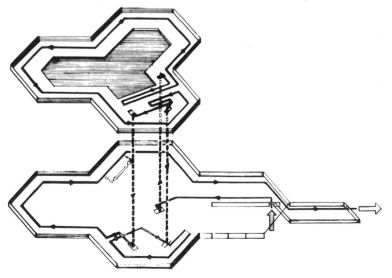

with a set of one- and two-story interlocked hexagons that permit each working group to have its own piece of real estate—generally one wall of a hexagon—and to remain visually and socially connected with nearby work groups. The corners of the hexagons, between work groups, are usually buffer zones for three carriers, so the workers in a group can choose their own pace as long as they meet the agreed daily production quotas.

Car assembly is, by its nature, a noisy process, because it involves a lot of pneumatic tools. One design goal was to reach a noise level that allowed normal conversations—about 65 decibels. We were able to achieve this in the majority of the production areas. Some work stations near noise-making equipment that is essential to car assembly had to be content with 80–85 decibels—still below the requirement for hearing protection. To achieve these levels, we used all the traditional measures and materials for cutting noise. In addition, we explored the noise sources themselves. Wherever we could, we chose assembly hand tools that give high output at low speed, which results in marked noise reduction. Then we added walls and screens with exceptionally good sound-absorbing qualities as well as a roof absorber system. The result has been a relatively pleasant working environment from an acoustic viewpoint.

The work that people do can be pleasant or unpleasant according to the design of the plant. Auto companies have done some work already on ergonomics, but most of this work is pointed towards more efficient and faster working rather than more comfort for the worker. In planning Kalmar and our other new plants, we have had industrial doctors involved from the start, so assembly work is carried out with the least physical strain and the greatest comfort. Orthopedic specialists help us design the best possible products; similarly, they have helped us develop comfortable work stations. We believe both investments pay off handsomely. The carriers have had similar attention, so they permit people to work in physically correct positions throughout the assembly process.

Building the atmosphere of the small workshop into a larger plant was a challenge. Most of the group working areas are light and airy, located mainly along the outer walls where they have large windows looking out over the countryside. Most of the twenty-five teams have their own work areas, their own entryways, and their own nonwork areas too. These consist of carpeted coffee areas and countryside rooms, each with a stove, refrigerator, coffee-maker, pantry and views of work. In addition, each set of two groups shares not only a foreman and production engineer, but also a sauna, changing rooms, showers, bathrooms, and closets for clothes.

The light, airy character of the Kalmar plant depends to a large extent on keeping materials in the center and only moving small quantities at a time to the team work stations. To keep the central stores within control, the entire plant has been working with only a two-day supply of most materials on hand. This is possible because all the systems by which Kalmar is fed work with similar effectiveness. It is possible to store larger quantities at Kalmar, but this would make the working conditions more crowded and thus less pleasant—and it isn't necessary.

The office building, another hexagon adjacent to the factory buildings, has two enclosed, landscaped courtyards as well as reception facilities, a waiting room, an assembly hall, lecture and conference rooms, and a health center. Although each team has its own relaxing area, we felt the entire work force should be able to establish contact easily, so the office building also holds the canteen for all employees.

## THE GROUP

The basis of the Kalmar concept is not the carrier but the group. The carrier was simply a means of making group work possible, practical, and pleasant. Today about 500 production people, organized in about twenty-five groups, are involved with each car. Sorting the various miniscule tasks that evolved in the technical era into reasonable segments was the prerequisite for what we might call the social era in production. These groups can view themselves as experts—in electrical systems, in instrumentation, in steering and controls, in finish, in interiors, and so

*Each group has clearly identified responsibilities.*

on. When a team has its own area on the shop floor and its own rest area, the ownership of that area seems to enhance the sense of membership in the team.

How the team organizes itself is up to its members. This is not spelled out by top management, nor is it the responsibility of the foremen, who act more as consultants and teachers now, with one foreman for two work groups in most cases. The carriers give the teams considerable choice in how they do the work. Most of the workers have chosen to learn more than one small job, and the individual increase in skills gives the team added flexibility, too.

Teams mainly choose to work in one of two ways. The first method somewhat resembles traditional assembly. The cars move on their carriers through the team zone, while each member does a different job on each car. The main difference from traditional assembly is that the team controls the pace, and the carrier normally stops unless a worker wants to travel while working. In the other mode of working, which seems to be growing more common at Kalmar, a team divides itself into subgroups of two or three people, and a subgroup carries out all team jobs on a particular buffer zone, and when they finish it goes directly to the outgoing buffer zone.

The only contract the workers have with management is to deliver a certain number of finished doors, or installed brake systems, or interiors, every day. If a team fills its buffer zones, the members feel free to stop for coffee or a cigarette. People often use this time to make minor adjustments or double-check their work. Most of the teams have chosen to work together, rather than in the line mode that simulates the old assembly line. Many people were interested in learning new tasks from each other.

To gain a sense of identification with the work, teams must also take responsibility for their work. Therefore, the myriad inspection stations that characterize most factories have been broken down at Kalmar. Instead, each team conducts its own inspection. After about three work stations, there is a special inspection station which has test equipment and people with special training. Three of these test stations late in the assembly cycle have rolling-road machines where cars are test driven under different road conditions. The computer-based system for quality information takes reports from these stations and flashes results onto the TV screens of terminals at each group station if there are any persistent or recurring problems. The computer also stores and feeds back information about how that group solved similar problems before. Perhaps even more important than the fault information is the fact that the computer informs the teams when their work has been particularly problem-free. The computer is not monitoring their work but helping them to do it better. While performing this task, the computer is also collecting the overall inspection information about each car as it goes through the line, so an inspection printout at the final station gives a sound guarantee of that car's quality and an historical record if the car ever encounters service problems later.

## THE RESULTS

To give people the flexibility to reorganize themselves, Volvo's production technology had to be changed. The purpose was to make the working situation more satisfactory and productive, but the initial thrust at Kalmar, and later at our other plants, has simply been to make the technical systems work. Kalmar involved a great deal of innovation and thus a great deal of risk. The carriers gave us minor problems at first. Another minor but irksome problem concerned the conductive tapes in the floors. Tidy people involved in the layout invented a machine that made little grooves in the floor so the tape was hidden under a thin layer of hard plastic. This gave the impression that it couldn't be altered without tearing up the floor—a physchological barrier to the flexibility we wanted. Eventually we got the little bugs out of the carriers and eliminated the resistance to regrooving, but it took time and attention.

People often ask why we concentrated so much on the technical systems and so little on the people for whose convenience they were designed, especially at first. The answer is fairly simple. If we failed with these technical systems, the chances of changing the work organization would drop drastically. If we succeed with the technical systems, we gain a visible, economic success, which

*Electrical tape, grooved into the floor, powers the carriers.*

is the prerequisite for acceptable overall performance. We could not "succeed" with the people themselves until we succeeded with the technology for people. As a result, we hope people will be motivated to reorganize their own work still further to suit themselves—and they will do a better and more dynamic job for us in the process.

Technology can strangle people. On the other hand, if it is designed for people, technology can also be a liberator. My hope was and is that the Kalmar solution would really be a tool for the employees, and that once its credibility was established, they would like to use it.

The best we can hope for, in Kalmar and our other plants, is to achieve situation in which technology does not limit the freedom of the men and women who work there. Then we may experience a dynamic kind of organization development that comes not from management but from the work force itself. An organization that develops and changes at the instigation of its members rather than its managers has a better chance of renewing itself all the time, evolving to fit the true situation of its people.

There have been critics along the way and criticism will continue because this is a new idea and it challenges tradition. Many said that people don't really want to be given more responsibility, and we would eventually have to go back to more standard methods. This has proven to be untrue. At one point, when we needed about forty people willing to learn new tasks to serve as a flying squad of relief personnel, we had over seventy volunteers—more than 15 percent of Kalmar's work force at the time. Extra pay for the increased responsibility was negligible. The thing that motivated the volunteers was variety. At the same time, rotation is not forced on anyone. Choices are available and it is simply up to the individuals whether or not they care to take advantage of them.

My original thinking about Kalmar was based on the idea that employees could develop work organizations to suit themselves. The importance of employee consultation became obvious a little later. Beyond the normal concept of management we can now begin to imagine changes triggered by the employees themselves. The finest form of organizational development occurs when planning can be entrusted to a group of people, ideally those who do the work.

Three phases can be identified in a development such as Kalmar, and they require different styles of management. In the first phase, it is the project leader who makes the technology work. If he or she fails, nothing has been achieved. In phase two, the employees become familiar with the technology and adopt it as their own. In phase three, the employees begin to take responsibility for parts of the operations, or the entire operations, and to exercise their own initiatives. Kalmar is not there yet.

I think it wrong to introduce changes in work organization as if you were making a big contribution to improving the lives of your workers. People are not generally enthusiastic about working with new technologies. They are suspicious. They seldom show gratitude for your efforts. I regard such a venture as quite successful if the "noise level" gets higher. That increase in itself reflects a constructive activity that offers promise and potential. Whenever we have been seen as preachers, pushing a doctrine that is good for others, we have suffered a boomerang effect. If our efforts to change the technical systems appear patronizing, the whole issue could die an untimely death. The suggestions and solutions for change must ultimately come from the workers, not the managers. It was management's responsibility to create a more flexible environment to begin with. It is the worker's responsibility to decide how to use it. The results of their decisions have so far been encouraging.

Kalmar's productivity fluctuates, like that of any other plant, but a definite upward trend is evident. As far as a cost justification goes, we are convinced it has been worthwhile. The initial investment was 10 percent above the traditional plan, mainly for development of the carriers, which have in themselves been a good investment. But this additional cost was so marginal that it would be hard to allocate it to different departments. The important thing is that the technology works. The plant is functioning. There is no reason why Kalmar will not continue to produce cars in successful competition with our larger assembly plants.

The assumption when we started was that the productivity of Kalmar could equal that of any comparable traditional plant. This has proved to be true. More important, this kind of plant has no built-in limits to increased productivity. Thus, we can hope for greater productivity in the future.

*The less physically demanding Kalmar assembly system encourages women to seek jobs in the plant.*

In October 1976, the results of the first outside evaluation of Kalmar were released. This evaluation was requested by employees at Kalmar, who thought the assignment should be given to The Swedish Productivity Council (Rationaliseringsrådet), a research body sponsored jointly by the blue-collar unions and the employers' association. The report was approved separately by the two sponsors, as well as the research body.

This evaluation was based partly on interviews with over one hundred blue- and white-collar workers at the factory, and partly on the observations of the research team. The findings are quite positive:

- *Nine out of ten workers participate in job rotation, and eight out of nine think it is a good way of working.*

- *The workers are somewhat free to take breaks because of the buffers, but the original goals have not been met.*

- *Decisions are delegated beyond the norm in traditional factories.*

- *Nine out of ten workers want to take responsibility for the quality of the product, and feel that they are partly able to do so.*

- *Final adjustments take more time than original plans anticipated.*

- *The employees say the physical working environment is very good. Original goals for noise, light, and health service have been met, but they asked for certain changes regarding working postures and summer ventilation.*

- *Productivity is as good as Torslanda, and the flexible layout will pay for the higher investment when it is used to full capacity.*

- *Personnel turnover and absenteeism are around five percentage points lower than Torslanda.*

| | Blue-collar % satisfied | White-collar % satisfied |
|---|---|---|
| Physical workload | 83 | 98 |
| Work positions | 55 | 70 |
| Noise | 80 | 97 |
| Lighting | 86 | 94 |
| Windows, outlook | 71 | 82 |
| Climate, air pollution | 49 | 76 |
| Protection against chemicals, etc. | 58 | 82 |
| Personnel space | 71 | 91 |
| Emergency risks | 81 | 91 |
| Medical care | 96 | 97 |
| Safety precautions | 86 | 97 |
| Workplace and environment | 83 | 94 |

The above table shows the reactions of sixty-nine blue-collar workers and thirty-three white-collar workers interviewed about various aspects of the Kalmar plant.

## THE SOCIAL CALCULATION MODEL

Rather than indulge in the numbers game, I would prefer to leave the Kalmar justification in the broadest terms—because its payoff to us is in equally broad and unquantifiable terms. Kalmar has accelerated the rate of change at other Volvo plants, for example. It has helped create for employees a sense of membership in something a little bit special, a trifle pioneering. It has added to our momentum. In general terms, for example, Volvo moved up from eightieth on *Fortune's* international 500 list in 1975 to sixty-first in 1976—in a recession year and a recession-vulnerable industry.

Rather than measure by traditional indicators, we tried to develop new tools to relate quantitative and qualitative aspects. Too often in planning, managers pay attention only to the factors on which they can hang numbers, leaving aside others—often much more important—which are harder to measure. Morale, employee attitudes, and momentum are difficult to take into account when making hard-nosed investment decisions. Volvo has tried to solve this problem with a "social calculation model" that goes beyond strictly economic terms to provide a new basis to justify investment. This takes into account assumptions about human behavior that are just as important as (if not more important than) assumptions about economic results. This model is now a standard tool that has been worked out in detail. The results from Kalmar have given the confidence needed to continue developing the model.

When we started thinking about the cost to the company, and the cost to society, of high employee turnover and absenteeism, we tried to derive from our own experience some statistics. The company has to bear the costs of recruiting and training employees. The absenteeism and turnover rates also increase the costs for quality control, for maintaining larger buffer stocks of semifurnished goods and components, and for adjustments, tools, and machinery. Administrative costs go up when a company must maintain pools of reserve labor to fill the requirements during peak periods of absenteeism. If the costs of high labor turnover and high absenteeism could be reduced, the resulting savings could be used to improve both the corporate and social environments.

Volvo consulted with social scientists in universities and public agencies as well as other specialists. Preliminary studies showed that the total savings, to a company and to society, would be substantial. A company with 10,000 employees would save about $12 million a year for itself and another $8 million for the community and country if it could reduce labor turnover by 10 percent and achieve a simultaneous drop of 4 percent in absenteeism.

This model made it possible to further justify investments in property, plant, and equipment when the improvements would improve job satisfaction. The Kalmar plant is one example. Similar calculations provided the basis for designing the Skövde and Köping plants. But the investments might not have been effective if we had not designed the new, improved work environments in close cooperation with employee representatives.

It was necessary to quantify these savings and build them into the same framework as our normal accounting system before this approach could become generally acceptable to the company. Today managers present a social calculation estimate—the improvement they expect mainly in labor turnover—along with normal cost estimates, for new investment. The social estimates, now based on a few year's experience, are usually quite accurate. Like other ideas that evolved during the Kalmar experience, the social calculation model is now a normal facet of management in every Volvo facility.

The real results of the Kalmar risk are now clear: there are real alternatives to the traditional assembly line, and that they do not need to increase production cost. Kalmar demonstrates that a less central role for production engineers and foremen does no harm to the quality or quantity of production. The accomplishments at Kalmar have helped to spark more general debate about the assumptions underlying traditional production technology.

Pioneering always carries a certain cost. Kalmar has attracted attention in other countries as well as in Sweden, so the first several years were marked by a stream of visitors to the plant. Kalmar's productivity may illustrate the so-called Hawthorne Effect, if the workers are motivated to some extent by the ongoing attention.

The start-up years also entailed getting technical systems to work and training many people in completely new types of organization. Now the start-up problems are behind us and the technical and economic objectives have been met. Advances in the future will depend on the real involvement of the workers and the changes they consider important.

Volvo has received many requests from social scientists and research people to make studies at Kalmar. However, their research was not very helpful when we were designing the plant, so we have resisted most of these outside requests to do attitude surveys and questionnaires. A few of our own researchers and those of the unions have made useful contributions. We don't want to find out what we did yesterday. We want to know what to do tomorrow. Changes in social research are now called for to match some of today's changes in society, so studies in the future will be less concerned with "proving" things that have happened in the past. Business is too complex for single factors to be isolated and generalized in the way that most current research tries to do. Kalmar met the company's desired objectives because of the people rather than the carriers, the hexagons, or the coffee corners.

## THE CARRIER OF THE FUTURE

The carrier was instrumental, of course, in making it possible to change the production techniques at Kalmar. It is still evolving and can be potentially extended in a number of different directions.

Carrier development has proceeded in two main directions: wheeled carriers and air-cushion devices. The air-cushion principle is not new: the device is very simple and a single person can move a ten-ton truck with a gentle push. Volvo's contribution has been to view it as a way to improve group-working possibilities, rather than simply as a device to move ten-ton trucks. Like the Kalmar carrier, it liberates people from the line. We are now able to use air-cushion devices for team working at Eskilstuna for tractors, at Arendal for trucks, at Borås for buses, at a new engine factory in Vara, and in the body shop at Torslanda.

The automatic carriers on wheels, orginated for Kalmar, have become the main product for a group called Volvo Engineering. They have simplified the original carrier computer system and reduced its cost. The wheeled carrier has evolved along two main lines.

The first is the Kalmar-type production platform, over which workers have considerable control. The second evolution is for pure goods transportation, with computer control permitting operation without human intervention. Production-platform type carriers are now being developed or used at six Volvo plants. Automatic material-handling systems have eliminated heavy or repetitive jobs moving pallets at the Arendal truck factory, and moving cars and bodies at Torslanda. One of the side effects of the Kalmar development was recognizing the information-gathering possibilities when such a device used with a computer system reports its position through simple sensors along it path. This idea

*Computer terminals reroute carriers and report quality status.*

may be useful in inventory control, or may have many distribution and transportation applications in the future.

We are not the only automotive company to use a carrier concept. Saab developed a small carrier for engine production in 1973. Fiat has two different projects using carrier concepts, and uses a carrier for making the model 131 Mirafiore car. Peugeot uses a carrier-type device to move body parts from an assembly group to the line. International Harvester in the United States is a pioneer in use of the air cushion. Rockwell International in Detroit uses the carrier concept for rear axle production.

The carrier idea should not be confined, of course, to the automotive field. Several firms already use carrier-type transport devices to eliminate unpleasant or dangerous jobs. Frigoscandia, which has 200 cold storage plants in Europe, is working with Volvo Engineering to develop an automated warehouse for frozen food. The unmanned, computer-controlled carrier not only saves people from working in below-zero temperatures, it can also save up to 20 percent of the energy cost because it works in the dark. Another joint venture with ARLA, Sweden's dairy production cooperative, also eliminates working in the cold. This is an integrated plan for materials handling, in which one kind of simple carrier feeds milk-carton packing machines with the paper they need to make cartons. A different kind of carrier with three forks takes three loads of full cartons and delivers them to 4°C cold storage. This carrier also picks up cartons from the storage section and delivers them to the trucking bay—all without human intervention. This final step involves our other big product line, trucks, so it may be possible to design future trucks for easier loading and unloading by carriers.

Eventually, combining the computer, the carrier, and trucks, it may be possible for machines to spend the night analyzing the day's orders, then planning and scheduling the packing machines, supplies, paper, and milk in storage, moving items around as necessary to get ready for the next day's business.

Other applications under discussion include a carrier model that can load at heights up to thirty feet in an automatic warehouse, one for cargo handling at a Swedish airport, and another for baggage handling at Munich airport. An airport of the future could use carriers and computers to move passengers, cars, cargo, and baggage in smooth progression between points that are bottlenecks today such as passenger terminals, baggage collection areas, garages, warehouses and the planes. Then a passenger might drive to the door of the terminal, leave his car and baggage there as he checks in, and be transported comfortably to his plane. On his return, carriers could have his car waiting for him and the baggage beside it. Carriers will be cheaper and more efficient than today's combination of trucks, buses, conveyor belts, moving sidewalks, and escalators to move people and goods through airports to their destinations.

Whether the carrier idea is used in an assembly line or a city center, the point is that the carrier lets you organize yourself as you wish and as the situation demands. For Volvo, the achievement of Kalmar is clear. Whether or not we have yet managed to organize the plant so people are satisfied with their work, we have certainly solved the materials handling problem that created so many people problems in the past.

The carrier is not a great invention. Automatic warehouses already exist. So do the independent transportation machines most of us have in our garages. The important thing is to view the transportation machine or production platform flexibly. Most automatic warehouses assume the user will reorganize hardware, production flow, and people to suit the system. The carrier idea presupposes that you fit the system to your convenience, your methods, and your people.

# 4

# Torslanda: breaking down bigness

almar is a pleasant town with a rich history, in a relatively flat area where farming predominates. We were able to handpick a skilled work force to start the Kalmar factory because we went into operation at a time when another factory was cutting its work force. Few foreign workers are included in the work force, simply because the area has few immigrants. One structural point that has contributed to our progress at Kalmar deserves emphasis: the plant was designed for a maximum of 600 workers per shift.

Change is quite a different matter in a place like our 8000-employee Torslanda factory, just outside Gothenburg. The city is busy and business oriented, characterized by granite outcroppings, trees, and parks. Gothenburg is highly industrialized but dominated by recession-prone shipbuilding and construction industries, so it suffers greater economic ups and downs than most cities. It is more a melting pot of populations than most parts of Sweden. About half the employees at Torslanda come from foreign countries, mainly Finland and Yugoslavia. The factory suffers high normal absenteeism related to its bigness and urban location, but due even more to the unique Swedish sick-pay system that gives an absentee 90 percent of his or her earned income. This helped boost the absenteeism rate.

Built in 1964, at the zenith of the technical era, Torslanda is tied to the assembly line on a large scale. Tearing it down and starting over is not economically feasible. New factories can sometimes be built on a smaller scale, with flexibility as a more important factor than economy of scale, but Torslanda remains the source of our "bread-and-butter" production, the mother factory for the cars that are the company's primary product. This means, too, that we can't make sweeping changes that might disrupt production, as we might be able to do in a new place where we were adding or replacing capacity.

Critics say: "Yes, Kalmar is nice, but most people can't afford brand new factories. So what do you do when you already have a big plant built around the traditional technology?" We think the people at Torslanda have developed some answers to these questions.

While central management spurred the Kalmar plan to a large extent, most of the changes at Torslanda in the past five years have been locally generated. For this we must give considerable credit not only to the managers there but also to the fifty-five works council groups and advisory groups, to a number of active union officials, and to the more than 8000 workers. Many of them have welcomed changes.

## A CONSULTATION STRUCTURE

Throughout the company, we have a hierarchy of works councils, with representatives from both management and the employees. Some of these councils—one per plant—have been required by collective agreement since 1946. Others, like the Corporate Works Council, have been created on a voluntary basis to meet our own needs for consultation.

The Corporate Works Council has been useful for both management and the employees. It meets about four times a year in different places related to the auto industry. One time the council will meet at the plant in Torslanda; another time the group might visit a dealer, or a remote plant, or one of Volvo's outside suppliers. We try to tie our meetings to operations related to the company. Once, after the council had backed the idea of the United States factory, we went to the United States for a week and visited other auto manufacturers, dealers, and places related to the auto business in America.

The Corporate Works Council is organized like most of the others in the company. The president chairs it and one of the employee representatives is vice chairman. Ten of the nineteen members are selected by the employees and the remaining nine are management people. Anybody can bring to the agenda anything that seems important. The works council is more of a consultation body than a decision-making body. Thus, we discuss investment often but do not vote on it.

The Corporate Works Council does have certain decision-making powers and a considerable budget to allocate. Under normal circumstances, works councils don't have money to spend for specific items, but our Corporate Works Council spends a good deal of money every year to improve the working environment; this is in addition to the budget of each unit for its own environment—another set of decisions in which the local works councils take part.

A special working group within the Corporate Works Council has been asked to set up priorities according to need in the various Volvo locations throughout Sweden, so the council members have visited every plant to help decide how to distribute the money. The corporate council has also designed our new personnel policy. Volvo is one of the few companies in Sweden in which both management and labor worked together on such a task and eventually achieved consensus on both content and implementation.

In spite of the effective consultation structure in the hierarchy of works councils throughout Torslanda, the plant's organization was and still is somewhat monolithic. Torslanda has large departments and considerable interdependence between departments, which often gives rise to friction or competition between them as well. Many jobs are still paced by the speed of the assembly line—a matter determined by negotiations between union and company representatives. The speed is monitored by computers.

Torslanda's employees actually have a strong say in the speed of the line, but, because it is negotiated through formal processes, the individual must see his or her own influence as less direct than the control a Kalmar employee exhibits. Three times a day at Torslanda the line stops; twice for coffee breaks and once for lunch. Stopping the line ostensibly costs the company money, but the employees asked for it so they could have their breaks together; we get better coordination and no real loss in overall productivity from the shared breaks.

Some kinds of inflexibility are built into this kind of monolith. Although we have flexibile working hours in offices and labs at Torslanda, we still have shift work and to-the-minute work schedules for most of the production people. Thus the early shift comes to work at 5:54 A.M., for example, and departs with the whistle at 2:36 in the afternoon. The only way we can erode this kind of clock control is to speed up the introduction of group working, which can give group members a little more flexibility in arranging their own schedules.

In an environment like this it is difficult to engender a sense of trust and membership in the overall unit. Smaller factories have more success in this respect. One factor which seems to help create a sense of membership in the smaller units we are trying to create within Torslanda is the information various departments give employees about their own output related to company-wide results. Overall, Torslanda turns out a competent house journal, in three languages. At the end of the year the treasurer and I make a slide film for employees explaining the financial performance. This is also published in a clear and jargon-free form in the house journal.

Since April 1975, by agreement at the national level between the unions and the employers' confederation in Sweden, all employees are entitled to full financial information about a company. They are also entitled to have a financial committee of their own, and a consultant if they feel they need help interpreting the information they want about the company and its operations. In Volvo's case, this created no problems because the company had already been working towards open information for employees for a number of years. The works councils agreed that our existing information was sufficiently clear and comprehensive that no outside consultants have been necessary. The financial information agreement has helped make more managers aware of the importance of good information clearly presented, and the specific feedback employees get within their departments is gradually being enriched as a result.

## ROTATION AND BEYOND

The main plant at Torslanda is almost a mile long, a symbol of the long lines that prevailed in auto manufacture when the plant was built. A car goes through four main processes before it is finished: pressing, body work, painting, and assembly. Each of these four main processes forms the focus of a major department at Torslanda.

*Torslanda stretches almost a mile from end to end.*

Organizationally, we would like to make the four departments as autonomous as possible. Thus, each now has its own decor. The body-shop changes in 1973 indicate how this minor item in the environment can contribute to a sense of participation that gives impetus to other projects as well.

Bedevilled by an inherently noisy and dirty process, body-shop workers and managers got together and chose a working group to assess the various problems, suggest some solutions, and figure out the costs of the alternatives. The working group went to the Gothenburg School of Applied Art for help. A group of architects there took on the shop as an assignment.

The results were promising. The architects proposed a fundamental color scheme with red, orange, blue, and green. They also suggested ways to cut the noise from jigs and grinding machines. These suggestions were put together in a special exhibition in the shop, and employees came to see and decide for themselves. Many had further suggestions and related problems to offer. The response was quite positive. It took several years to implement all the proposals, but today the body shop is one of the brightest spots in the corporation. By building "houses" for the worst noise-making equipment, the designers were able to cut the sound level by 10–20 decibels in the noisest areas; new equipment keeps the air cleaner where dust was worst. The pleasant surroundings encourage people to keep the area tidy, too.

Although the consultation structure has always been strong at Torslanda, real changes began to come to light as Kalmar planning got under way. Torslanda was in some ways the cradle for Kalmar. The carrier was developed by people borrowed from Torslanda; then, when it came time to try out the tilting device for the Kalmar carrier, a number of tilters were installed in the assembly area at Torslanda, replacing pits. The results were so positive that twenty of them are used in Torslanda assembly today. The Kalmar project acted as a catalyst, as we had hoped, heightening the awareness of other employees and managers, and spurring creative ideas about alternative work styles.

Gradual change was the goal at Torslanda, and this has been achieved. Two parallel strategies were developed for the big factory. First was to mechanize wherever it was feasible to do away with dirty, heavy, noisy jobs. Second was to change the job where it couldn't be automated. Gradual improvements along four fronts can be noted. The first is the physical environment, primarily improving the working place itself, with attention to the secondary places such as green areas, recreation, lunchrooms, and so on. The second front is job design. The third is participation, including the committee structure that constantly changes and grows, setting priorities for improvements and investments, and making decisions regarding working patterns.

Fourth, and increasingly important as we make progress on the other three fronts, is personal development, the chance for individuals to learn more, to enhance their personal lives and careers through opportunities available within the company. Whether people are career-minded or not, they have chances now to learn what we do and how we do it, as well as to gain any specific knowledge they want.

With improved motivation and involvement, people at Torslanda are no longer "assemblers." In recruiting, we at Volvo now use the term "car builders." This is part of an overall drive to achieve higher status for the people who have already begun to contribute more than mere muscle power. Changing the term may help change the image, but it cannot be a gimmick; this kind of approach only works when it represents a real change in the work itself as well as in attitudes.

Starting change and keeping it going is a challenge in a place like Torslanda. Most of the works councils, consultation groups, and project groups have money to spend on such things as improving their working conditions. Changes to cut dust or noise require little outside impetus. Changes of work structure that can take place within a single working group also occur easily. But we have discovered that most real changes of work structure affect other groups, at least indirectly. So the problem has been to draw the boundaries for single working groups wherever we could, and to make them as clear-cut as possible.

Change is more difficult when it involves the organization itself—and that includes identifying and separating working groups. It also includes anything that crosses departmental boundaries, because the need to coordinate between groups and

*First changes in paint and body shops concerned physical comfort.*

*Wherever possible, jobs like welding have been completely automated.*

the number of interested parties escalates rapidly. There are always a number of practical reasons to maintain the status quo. The more organizations one finds involved in a change, the more reasons will be put forward to retard change. Only when a clear, corporate-wide drive for change is visible and credible will resistance points subside. In this context, once you have some easily identifiable working groups, job rotation is fairly easy; but job enrichment usually crosses boundaries, of status if not function or organization, so it often takes longer to get started. Enrichment seems to be worth the trouble once it gains momentum, though.

The basis for today's progress came from some steps that were taken in the late sixties. In the mid-sixties we experienced a very rapid increase in production, which meant recruiting at least 1000 new employees every year in addition to the normal number needed for replacement. As absenteeism and employee turnover rose, the recruitment problem grew worse. In 1968 managers and union representatives started a discussion group and employee interviews especially about absenteeism. That group and the study it conducted created the groundwork for later advances. One recommendation the group made then was for job rotation. Volvo has that now. Another was providing more interpreters and other social services for foreign workers. We have those now, in addition to language lessons on company time. The third area they discussed was the job environment. That has taken the most money, but progress is visible and steady. In a progression that would be familiar to Maslow, we find today that personal development and a chance to influence the work have become more important to the employees than they were a few years ago.

The approach at Torslanda is evolution—step-by-step improvement, with people involved every step of the way. We can't measure the pace, but we have progress every year. Every summer when the company shuts down for vacation, different areas are improved. The employees have been involved in almost all the changes—ventilation, lighting, layout, noise, dust, decor. That's the hardware side of the improvement. The software side is job design.

Job rotation was the opening wedge. It started around 1964 in the upholstery department, for very practical reasons. Employees complained of sore muscles from doing the same operation over and over. They discovered that if they paired off and traded jobs every day or so, they were able to use different sets of muscles. From that ergonomic, down-to-earth beginning have grown most of the other changes that focus on the quality of work life more generally. Soon the upholsterers (mainly women) were trading among three or four, and the relief from wrist, arm, shoulder and back pains was far greater than anything they had been able to get from the small army of supervisors, doctors, safety specialists, and industrial engineers who had developed methods or mechanical aids to relieve their complaints. There were disadvantages, though some workers found the change of pace from one job to another too difficult, and wanted to go back to the old system.

There were some benefits. Most of the employees felt they understood the overall process a little better, and had more tolerance for their colleagues when they learned that the other jobs were not as easy as they thought. Improved contact with fellow workers sometimes led one person spontaneously to help another. The physical pains (and associated absenteeism) dwindled. But feelings remained mixed.

In 1966 new equipment changed the nature of the work, and forced a two-month period of one-station work with slightly more complicated tasks. Then production engineers, in consultation with upholsterers, gradually developed a completely new system in which each worker learned the work for all fifteen stations— most of them about two-minute tasks. Then workers rotated every day or half-day. Seats with errors were immediately sent back to the proper station to be fixed. Each operator had training to learn every station, so there was more variety. They noticed a sharp drop in the aches and pains, and new signs of team spirit began to show up. This time the workers were overwhelmingly in favor of the change.

Eventually the workers in upholstery were given responsibility for planning all their own work, a fairly complex task because the blue seat with the special fabric has to leave the upholstery group in time to reach the assembly line at the right moment to meet the blue car with the special headliner. The planning had been a white-collar job in the past, but the group members, who alternated the paperwork task, had no trouble absorbing the planning. They get information on the assembly requirements from a teleprinter, and have to translate this into individual assignments. They have also taken over checking the quality of incoming upholstery materials, a job that used to require four inspectors. All the operators attended a short training program to learn how to find and identify all the kinds of defects and what to do about them.

As these changes took place, rather gradually, there was a concomitant increase in team spirit and individual commitment. Employee turnover and absenteeism changed dramatically, and the upholstery quality improved because team members understood the entire process and felt much more responsible for the product of their work.

The gradual growth of this group, from the first 1964 rotations to the full-scale enrichment and group working it achieved by 1968, has many of the characteristics of other successful changes. Initial resistance to change is overcome only when there has been sufficient time and contact between members for a real group to begin crystallizing. This contact is very difficult to achieve on a rapidly moving assembly line where task times are counted in seconds, not minutes. So the upholstery group demonstrated first of all that a group creates itself; it cannot be created by someone else. And the process does not take place overnight. Once a real group does exist, it can take on other tasks well beyond its original purpose. As later groups also learned, rotation was often better than a single short-cycle task, but it entailed a lot of hopping. If group members could enlarge jobs, taking in neighboring tasks, they could turn a two-minute station time into four, or eight, or ten minutes, and have more satisfaction in handing on a fairly finished product. Enrichment, whereby the group took on supervision and control and planning tasks, usually came later, when the group was quite well formed. And the enrichment gave further impetus to team spirit and morale.

Although the upholstery group provided a good example, and a plant-level study group concerned with job content was formed by 1969, rotation nonetheless crept on very slowly. By the end of 1970, about 3 percent of Volvo's 3000 assembly people were rotating jobs, although job-rotation missionaries in the plant and personnel people were pointing out its benefits frequently. In the 1971–73 period rotation absorbed a large quantity of planning and discussing energy for a rather slow visible result. We hit 10 percent in 1971, then 18 percent in 1972, then began to see momentum when one-third of the assembly people were rotating by the end of 1973.

From then on, there was a sharp rise in people's acceptance and willingness to try rotation. Less energy and planning were required from management to keep the momentum going. In 1974 almost half the assembly people rotated jobs, and in 1975 it was 54 percent. By the end of 1976 the number in job rotation, job enlargement, or job enrichment was over 60 percent and management knew it would continue until it gradually reached the natural demand for change in jobs. There are now and will to some extent always be a small percentage of workers, especially older people, who don't want to change at all. In one survey of 100 employees, 65 were already rotating and all liked it. Of the 35 who were not doing so, 23 wanted to rotate, and the other 12 were quite firm in their desire for the security of knowing they could come in and do the same job tomorrow.

The enrichment figures are still low but they are gathering momentum too. Experience in the paint shop shows the pattern. The grinding lines there, where bodies are polished after their base-coat painting, used to have operators who went over the entire surface with water-cooled sanding machines—an unpleasant job at best. Four controllers inspected the sanding and then sent any defects on to the elite adjustment polishers. The department had the familiar problems of absenteeism and turnover. A working group including supervisors and workers was formed to see how they could make jobs on the sanding lines more interesting and improve the quality. The group surveyed all workers, and ultimately suggested that the people who did the sanding should inspect their own work and decide whether supplementary polishing was needed. This, of course, meant that the polishers had to learn more about the entire painting process and how their quality was judged. After a long training course and on-the-job training, they were able to take over the inspection tasks; the polishers were the first to do their own inspection, but later the machine sanders followed the same pattern. By this time, all the workers could switch between overall sanding and spot polishing, and the group began to share responsibility for the quality of the bodies leaving the polishing line. The extra costs of training were quickly recovered; the four inspection jobs were abolished, and the number of defects leaving the shop dropped from an average 3.2 percent to only 2.5 percent. Absenteeism dropped to levels that were about the same as those in the rest of the paint operations, but personnel turnover was cut in half, while it remained about the same elsewhere.

A visit to this line reveals other differences that have evolved since the job enrichment project took place. Instead of a foreman, the group members have an "instructor" on hand in case they want advice. On one representative day, three people (two of them women) were hand-sanding the more intricate roof edges while five worked with machine-sanders on the sides of the car, and two more did the front and back. They had organized themselves this way. After coffee-break, the hand-sanders and machine-sanders traded. They had evolved from a line organization to a complete group working on one car, but were discussing instead working on three cars at a time, in groups of three, which would create less interference and splashing. The working arrangement was completely up to the group members.

## FOCUS ON GROUP WORKING

Group working is easier to achieve in body and paint shops, but Volvo is still largely tied to the line in assembly. One of the most important changes at Torslanda has been to move as much work as possible off the assembly line and into preassembly groups. This not only gives the group members some of the same social and work advantages the groups at Kalmar enjoy, it also makes the line itself a little less crowded, so it improves working conditions for everyone.

A reference group with four union representatives and five company representatives meets once a month to keep track of

*Torslanda is trying to increase the number of jobs suitable for women.*

progress in Torslanda, trying to analyze what is good and where emphasis should be shifted. The union people have initiated a strong drive to shift the focus from rotation to group working, setting up production groups with more autonomy. This makes a great deal of sense, because once groups exist and have autonomy, they can choose for themselves the rotation, enlargement, or enrichment patterns that best suit their own tasks. A four-man team from the reference group has set out goals and some guidelines for creating groups. An important element is that group working brings fundamental changes to the foreman's role. Instead of giving assignments, the foreman gives advice. It is noteworthy that the foremen's union representative was one of the four people who wrote the joint statement.

The joint statement identifies three vital factors for viable production groups:

- *There has to be some chance to identify group goals related to production, quality, or quantity, so members know what is expected of them.*

- *Some natural linkage should exist between the different tasks in the group, either the same product or some other natural communication. To get the kind of joint responsibility that marks a workable group, it is important that every member of the group knows, or can learn, most of the jobs of the others.*

- *A group needs good communication internally. Language barriers can cause problems. Therefore, with a mixed-language work force, this factor needs to be taken into account. Communication between groups is also important. Therefore, at least one member of each group should be fluent in Swedish.*

The degree of communication within a group also depends on whether it has a buffer system and can set its own timing.

Furthermore, we have learned through experience the importance of letting the group itself decide how its work is to be changed. Reorganization has to grow from inside. On the line situations will still arise where groups cannot fulfill all these requirements. Even so, thirty people could be reorganized into, say, three groups, and people could be shifted between groups to fit their capabilities and desires. Where we can't have buffers on the line, we are beginning to emphasize something learned at Kalmar. There the working groups didn't like having different relief people every time they needed a replacement for a member who was sick. So they created a pool of people who knew a number of tasks, and assigned them to specific sets of groups. Increasing the number of relief workers on the lines at Torslanda and making them specific to certain groups may help alleviate the absence of buffers, and make the relief people familiar and friendly semimembers of the working groups.

The benefits of groupworking are obvious, in terms of quality and cost for the company and better work and interrelationships for the workers. Groups have an effect on absenteeism, too, beyond the drop that naturally occurs when the work becomes more interesting. In one case in the body area, a man in the work group was absent one morning, so group members phoned him at home and said: "Why aren't you here?" He came in to work that afternoon.

There are problems associated with the change to group working, of course. Many outsiders wonder what happens if the group rejects a member, or how groups succeed at selecting their own members. We anticipated the rejection problem, and set up in-house employment agencies to find places elsewhere for people who had trouble getting along with a group, but there have been rather few instances of this. More often, the request for a change to work that seems more interesting to the individual.

One six-member group in the body area was growing to eight members. The group chose its new members. One of the newcomers was a pleasant person, but was absent quite often. The foreman did not approve of the choice, but said and did nothing. He found that the group, having made the selection, seemed to have a joint commitment to keeping the man at work.

*A small roof over the work area helps create a sense of being a separate group inside the giant factory.*

The company, the unions, and individual workers agree that a vital prerequisite for group working is that demands for production rest on the group, not on the individual members. Instead of measuring individual productivity (and paying individual bonuses) the company simply sets a norm for the group.

In any group, of course, differing work styles must be expected—some members do more than others. The members generally accept the differences gracefully and adapt to them. As long as the group attitude is supportive, small differences don't seem to matter. This leaves ample room for individuals to have a bad day or a good one, or to work better in the afternoons than in the mornings. Furthermore, it gives the union a better chance to help its members. If someone is upset about his or her job content and complains to the union, the union looks first at the group situation rather than at the individual's situation. The union people can often solve problems with the group rather than in isolation, and small events that might otherwise become politicized tend to stay in perspective.

An effort is being made in Sweden today to find jobs suitable for people with handicaps of various kinds. Group working gives us a good opportunity to do this. Disabled people travel from group to group evaluating the work and discussing with members which jobs might be done by handicapped people. Finding such jobs is specifically the task of the internal employment agencies.

Problems do occur in the transition from line working to group working. Both management and union people are particularly concerned to keep group working from taking on "elite" overtones. Instead of selecting only the most responsible people for one new group project as (management might have preferred) or a handpicked representative sample (as union researchers once suggested), Volvo went through the normal process of advertising the openings internally, listing as qualifications only that applicants should have one year with the company, some experience in final assembly, and must speak Swedish.

Over 150 people applied for the eighteen jobs. A random selection from the stack of applications resulted in a representative group; the blind selection process produced a group including several workers each from Finland and Yugoslavia, and four women.

Another group demonstrates a difficult but intriguing aspect of the elitism problem. An eight-member team in the body shop has normal shifts of 8.2 hours, like the rest of the plant. However, the team normally fulfills its daily quota within six hours or so. There is no point in trying to produce more because the team's task is part of a larger production schedule, and this would also upset the bonus balance with other workers. Members can't simply go home when they finish, because other workers are unable to do so. So, they tend to have coffee together, relax, use the recreation or sauna facilities, and have planning meetings. Sometimes they use the time to explore adjacent groups and see how the quality of their output is perceived by neighbors. This kind of situation will disappear when more of the other functions are done in working groups and there is less dependence on the whistle.

Interdepartment rivalries may be heightened by shifts to group work. The paintshop, for example, changed to group working in 1974, with workers doing their own inspection. A real improvement in quality has resulted. But there has also been an increase in complaints about quality from the department that receives the paintshop's output. The paint group considers the complaints arbitrary. It seems that when department A changes its work patterns, department B tends to look even more closely for flaws—an indication of intersibling rivalry which probably enhances group formation, although it is a nuisance. The problem seems to continue until or unless department B undergoes changes too.

## DOCK ASSEMBLY

The eighteen people chosen at random from 150 applicants were pioneers for a new kind of working to replace traditional assembly lines. They have two final assembly docks, with groups of nine people doing everything: chassis assembly, body work, mating the chassis and body, and doing the final trim and checkout. It takes an assembly group about an hour to make a car.

This fits the qualifications for our social calculation model—the production cost remains competitive, and the additional investment is justified on the basis of improved working conditions and content. In the new assembly docks, which went into full-scale operation at the end of 1976, the eighteen workers will be supported by more materials-handling people than before—five instead of two—but these jobs (fork-lift driving and so on) were already among the most desirable to employees because they include variety and social contact, so there is no job impoverishment involved in the change. Every working place has a buffer. The dock workers have production meetings for an hour or so every week to discuss their situation, which includes the usual technical problems that occur with new equipment.

The technical changes are expensive, especially at first. But dock assembly is moving in directions where real group working will eventually be possible for more of the line workers than ever before. We still have to move gradually at Torslanda, but significant changes have taken place over the past five years. Volvo wouldn't invest in dock assembly if the economic justification as well as the social justification were not apparent. We can already see that it is possible to cut production time sharply. When you split work in the normal MTM way, you find unused time between tasks for each worker. With larger job cycles, more work is done in total, and the work is more satisfying for the people. The nine people in the Torslanda dock are doing everything all the teams at Kalmar do. Kalmar itself was justified on the basis of Torslanda MTM figures, so the people involved in the dock project at Torslanda believe we have clear potential for more efficiency.

Management of group working differs markedly from traditional line management. Instead of giving orders, the manager has to listen, argue, motivate people, often compromise. This takes longer. Decisions are slower. But, in the long run, decisions are accepted and implemented rapidly once they are made.

We have learned something else from the group working at Torslanda. The success or failure of an idea is often attributable to whose idea it was, rather than any intrinsic goodness or badness of the idea. If it is the union's idea, or if it comes from a work group, an innovation has a good chance of succeeding.

If it is a management idea, its chances are slimmer. So, the function of management in a group-working context is not so much having ideas as creating an environment where the people who matter will be able to have ideas and try them out. Joint consultation on an informal basis—best exemplified in group working  gives a group a good chance of developing ownership of an idea and, therefore, encourage group members to make sure it succeeds.

# 5

# Building momentum

**V**olvo is not just Torslanda and Kalmar. The company has twenty factories in Sweden and seven abroad, in countries as diverse and distant as Iran, Peru, Malaysia, and the United States. Building a momentum of change toward more human working patterns in all these plants was a major objective from the moment Volvo began thinking about a nontraditional approach to the new Kalmar factory. In retrospect, the atmosphere of change toward participation was contagious, but the effects were confined to Sweden until fairly recently.

In almost every case cooperation and participation began with attention to the physical environment; only as that began to improve did people find it natural to turn their attention to the content of the work itself.

Over the past five years Volvo has invested around $20 million to improve the physical working environment for employees. This is simply part of the cost of achieving cleaner, more pleasant surroundings. It demonstrates in concrete, visible ways that the company values the people who work for Volvo. But this investment does not create better jobs. It only provides the conditions in which people can work together to organize their work in more human ways. So, attention to the physical environment is one initial step Volvo has observed in all its factories.

Another common denominator is the nature of Volvo managers who have encouraged the changes. They are usually hard working, pragmatic people, rather nonacademic in their approach. They tend to come from inside Volvo, and exhibit strong loyalty, both to the company and to their own divisions. Thus, there is limited interchange between truck and car divisions, for example. Many of these managers have engineering backgrounds, even those in such diverse jobs as personnel or finance, although there is a definite trend towards general management skills with a stronger social orientation.

A third factor that seems to permeate the Volvo experience in Sweden is the serious, mature attitude of union officials, who have often initiated or supported changes that caused them short-term inconvenience and required adjustments in such basic items as wages structures.

Group working, for example, means that the basis for incentive or bonus payments shifts from individual production to group production; yet it is the union, as much as the company, that has made group working a primary goal for Torslanda. The support of employees in general and their union officers in particular has been a positive factor in creating momentum for changes in work organization throughout Volvo.

Beyond these unifying factors, though, we find the differences between one plant and another are the pacing items for change. You can't transfer a design from one company to another, or even from one plant or department to another. Each has a different approach, a different culture, different people, a different set of problems, and different kinds of interconnection with employees, unions, and headquarters. A worker in our truck factory at Lundby, for example, has inherently complex tasks. A truck is more complicated than a car and moves through assembly more slowly. Three- to six-person groups do the assembly tasks, which results in high involvement with the finished product. The main change there has been to encourage groups to choose their own leaders and to rotate the leadership every month or so. At Torslanda, on the other hand, complete reorganization of the assembly process is our eventual goal.

If I had to emblazon two slogans on every wall of the company, they might be:

- *There are no experiments.*
- *There is no single solution.*

Within that framework, it may be useful to explore some of the individual situations that have led to advances in working conditions and work organization at different Volvo plants in Sweden.*

---

* Many of the changes mentioned in this chapter are described in greater detail in *The Volvo Report,* by Rolf Lindholm and Jan-Peder Norstedt. This study was published in 1975 by the Swedish Employers' Confederation, Box 16 120, S103 23 Stockholm.

## SKÖVDE: THE "E" PLANT

Like many companies, Volvo uses private shorthand to designate its various facilities. The factory at Skövde, a town of 35,000 in the southwest of Sweden, happens to be called "the E plant." It also happens to make engines. And now, by coincidence, it has a new building in the shape of a four-armed "E".

The engine factory at Skövde dates back to 1868, long before the birth of Volvo itself. It has a tradition of advanced industrial engineering and high productivity. Volvo was a good customer before it became owner of the engine company. Today the Skövde facility has about 4500 employees, a third of them foreigners (mainly from Finland), turning out about 300,000 gasoline and diesel engines a year.

Rapid growth of the company has been reflected in equally rapid growth of the engine division. The shift towards higher truck production led to a current project to expand the diesel area. The foundry was recently renovated at a cost of about $14 million. The main symbol of change at Skövde, though, is the E-shaped unit that went into operation in late 1974 to make Volvo's four-cylinder B 21 gasoline engines. Here we started with a new building, a new product, and, to a large extent, a new work force, because this was an expansion project rather than a revamping job. However, employees were involved from the first planning stages, and their contributions were important in the eventual success of the factory.*

Engine manufacture differs considerably from auto assembly. To make a cylinder block, for example, you have almost 500 separate cutting operations. The engine assembly process involves a comparatively small number of different parts, but requires intricate adjustments and tests. The machining is noisy and, before Skövde, technical advances had made the work rigid, with people as machine-tenders. The answer for the new Skövde fa-

---

* The degree of employee involvement is highlighted in a report by Noel M. Tichy, Graduate School of Business, Columbia University, entitled *Evaluating Organizational Innovations: Work Restructuring at Volvo and General Motors.*

*Skövde "E" plant.*

cility was to automate the machining as much as possible. Those jobs that remained involved overall control of, and responsibility for, the entire task. Assembly jobs could be done in quieter areas, in work groups. With Kalmar already under way, it was easy to think about work groups, carriers, and nonstandard architecture. The company/union planning group decided to have a building with four short legs for the four machining lines, with assembly in the large open area connecting the arms. A one-story corridor connects the other ends of the E, and the spaces between the arms contain gardens and rest areas. The most important benefit of this concept is the ability to work as five smaller factories rather than one large one. Each element has its own resources to manufacture its own product line. Workers understand the total process better, and they also have a better chance to get to know each other.

Skövde's plant manager was an individual who welcomed the return to a craft or trade orientation as a return to human-scale industry. He saw his role as helping employees learn from their own experiences, and creating an environment where they could be rewarded for trying new methods and encouraged to develop their own ability. His attitudes were made quite clear to other managers. They were in complete accord with the planning group's decision to make group working the basis for the new plant's organization.

The Skövde factory started up with a nucleus of experienced people and a rapid intake of new employees who tended to be young, well educated, and mainly Swedish. The ability to speak Swedish was a prerequisite for working in the new plant because so much would depend on communication among workers. There was also a drive to recruit women; new equipment in the assembly area reduced the heavy work to make these jobs more attractive to women. New workers usually started in the final inspection area so they could see the total product first. This helped put their later training in context.

The machining departments at Skövde are highly automated, with automatic loading for most machines, mechanized transport between operations, magazines for intermediate storage so machines don't have to wait for each other, and mechanical aids to

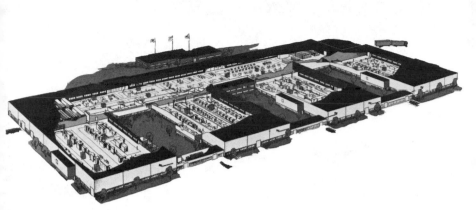

*Cutaway view shows how machining shops are like separate factories.*

get rid of heavy lifting. Now that operators are free from the monotonous task of feeding machines, they instead work in teams to supervise and control the machinery, cooperating to replace parts as necessary. Nobody is tied to a single machine. The workers have gradually built up a strong sense of team loyalty. They generally sort out assignments among themselves in a few minutes before the start of a shift. When one machine line stopped, an American observer was surprised to find that the workers, instead of taking a rest as he would have expected, all gathered around the culprit machine to discuss the cause and help repair it.

Skövde engine assembly takes place on a carrier analogous to Kalmar's but smaller and less computerized. Guided by the electronic tape in the floor, this little carrier holds the engine throughout assembly and test, with devices to move it and change the angle as necessary. The carrier moves from group to group under the control of the workers. Cylinder heads and valves, camshafts, rods, and manifolds are preassembled in separate groups. Inserting parts usually takes about three minutes at each station; however, because the workers know all the jobs, they can choose to work in teams to do a complete assembly job on one engine if they prefer.

*Small engine carriers at Skövde.*

Four to six people also work as final inspection and testing groups, running the completed engines under various conditions. They immediately feed back information to assembly groups if there are faults—or an extraordinary absence of faults. Buffer zones allow workers to control the pace of work, as well as its organization. There is no fixed schedule for job rotation—simply a flexibility that permits natural teamwork.

About two years after the new Skövde factory started up, we were able to evaluate it in general terms. The results have been quite satisfactory. The quality of the engines coming out of the E plant is consistently high. The quality of the work organization is also high, judging from the extremely low employee turnover and absenteeism rates. Targets for budget and production costs have not been met or exceeded.

Quantity improvements were never demanded, because over-production of engines would simply raise Volvo's work-in-progress costs, but we have found a levelling effect: if a tool change cuts productivity on one shift, the next shift usually makes up for it. The factory has been able to achieve the needed production with a smaller work force than expected. Even so, low absenteeism and high productivity give Volvo a slightly larger work force than the bare minimum. This encourages experimentation and gives people plenty of time to learn new jobs if they choose.

## UMEÅ: SAFETY FIRST

The nucleus for employee participation at Volvo was Umeå, a truck plant on the rural north coast with about 900 employees. In 1968 this plant experienced a short wildcat strike, and absenteeism and turnover rose sharply. Managers and local union leaders got together to review the situation and decided to gather opinions from the employees themselves at a series of evening meetings. The majority of the employees attended and contributed to the changes that were eventually made.

Umeå already had an effective works council, but small matters in particular work areas tended to occupy most of its time. Out of the evening meetings grew a structure of local participation groups.

## OLOFSTRÖM: AUTOMATING
## OR BUFFERING PROBLEM JOBS

One of our largest plants is the highly developed Olofström factory in the south, whose 500 employees make pressed and welded body components—dirty, noisy, smoky, and sometimes dangerous work. In the late sixties, all the usual symptoms of worker dissatisfaction were apparent at Olofström. The working environment was thus the first concern, and remains a primary focus for change.

Olofström set up a joint management/union committee in 1970 to see what could be done about working conditions. The committee selected four production areas with different problems as pilot projects; proposals for specific improvements were worked out with the foremen, production engineers, and workers in the affected areas. These first improvements rapidly led to discussions of job rotation, then job enlargement. Once work groups had been set up in the pilot areas and new equipment installed, the groups began to absorb some tasks formerly done by foremen and technical specialists.

Education for the "victims," people whose jobs were absorbed, eventually helped them enrich their own jobs. The management-development program for foremen at Olofström emphasized how to delegate to groups, how to help individuals in their personal development, and how to shift from authoritarian roles towards guidance and assistance roles. Technicians, especially industrial engineers, at Olofström had a similar situation. Instead of doing the job redesign, they now participate with workers in this task. They developed a higher view of their goals; the exceptional advance of the buffer concept in Volvo factories is an outgrowth of their work at Olofström. For the technicians, the change means that employees appreciate and use their special skills, instead of resenting their intrusion, as was sometimes the case in the past.

With so many inherently unpleasant jobs in pressing and welding, Olofström's approach has been to mechanize an entire work station where feasible in order to eliminate monotonous work. Where it is not practical to mechanize the job, they try to insert buffer inventories in the intervals between operations to cut the rigid link between the worker and a single task.

In the eighteen pressing lines, for example, one person used to be assigned to each of the five or six presses on the line, inserting a sheet in the press every few seconds. The workers checked and packed the output, while specialists maintained and adjusted the machines. Four of the lines are now completely automatic, so workers no longer need to feed machines and lift heavy stacks of sheets. This also improves safety, because the worker doesn't need to be near the press while it is in operation. Magazines at the head of each line hold about 400 sheets for automatic transport to the presses. The pressing work force has been halved on these lines, and remaining operators still inspect and pack the output, but they also adjust the presses and change the tools, the intricate jobs that used to belong to the specialists. Pressing output has been more than satisfactory.

Every factory operation has waiting times. The buffer concept doesn't eliminate these, but it shifts the waiting so it suits people rather than machines. We had dramatic demonstration of the benefits of buffers when we decided to put the buffers out of operation on Olofström's front fender department for three weeks. Production immediately dropped by about 10 percent. Furthermore, when researchers surveyed the workers—who had previously reported that the work was relatively free and flexible— they said the work seemed rigid and monotonous. The change in attitudes was so sharp that the workers asked the researchers to stop the test after the first week.

The investment in buffers, magazines, mechanical lifting aids and so on pays off handsomely. At Olofström we found the following results:

- *Profitability was maintained.*
- *Labor productivity was maintained.*
- *Employee turnover was cut to one-quarter the previous average.*
- *Absenteeism dropped to one-half the previous average.*

- *Recruitment became easier.*
- *A pool of more highly skilled workers was available.*
- *Quality improved.*

## BERGSLAGS: ORGANIC EVOLUTION

The Bergslags transmission component group is really four factories employing about 4000 people, with 2600 of them at the mother plant in Köping and daughter plants at Uppsala and Lindesberg, plus a foundry at Arvika. Work reorganization has enabled the role of the Köping factory itself to be upgraded, with new production of some marine engines for the Volvo Penta group, as well as mechanical gear boxes and axles.

In 1970, sharing the concern that was prevalent throughout Sweden at the time, the division manager set up a project group, including union representatives, to improve job satisfaction. The group's field of action covered the environment, personal development of employees, mechanization, changing work organization, and introducing new employees. Mechanization to eliminate environmental problems and improve short-cycle jobs, which were numerous, took center stage at first. The mechanization team set out to list and then eliminate the "ten worst" jobs in the factory. Workers were involved in every layout change during the mechanization. One result of this activity was development of a "mechanical hand" for jobs too complicated for industrial robots or simple hooks. These are cheaper, smarter, and more flexible than robots, and give the control to the operator instead of a computer.

In one flange production shop at Köping with short-cycle jobs, mechanization cost a little over $700,000 in 1971. Capacity then was 220,000 units a year—not quite enough to meet the company's needs. After new equipment speeded up feeding materials into the machines and between them, the population dropped from five machine operators to two who supervise the entire operation. Annual production went up to 475,000 units and manufacturing costs per flange dropped by about 25 percent.

A funny thing happened at Köping. Most of the management attention there has focused not on work changes but on informal discussions, creating a spirit of cooperation. Most of the work reorganization, as a result, seems to have flowed rather naturally and organically out from the shop floor. No one can quite explain why, but some workers who have been there a long time simply began to assume more responsibility.

There were no management demands for change. Workers just began to come up with ideas of changing jobs among themselves, even though they knew and cared very little about concepts such as "job rotation." In fact, one 1970 job-rotation scheme had been abandoned, mainly due to conservative attitudes among foremen. Now Köping people have not only developed some autonomous working groups, they have also changed working patterns completely in some areas. Instead of two nine-hour shifts per day, for example, workers in one part of the plant in 1974 decided to have three six-hour shifts—a pattern that makes it easier to attract local women into the workforce. Group members often work extra hours if one member has to be absent. Both the union and the company backed the employees in this change, in the face of some outside concern at nonstandard hours and changes in pay structure.

Management at Köping put its energy into finding ways to change attitudes. A series of Saturday meetings, for example, involved 95 percent of the white-collar and 70 percent of the blue-collar workers. These were informal, with discussion meetings during the day and a party with spouses present in the evening, at the company's expense. Employees also have vocational training programs, one-hour lectures to increase working knowledge, and they participate in planning new facilities. An annual day-long conference for all foremen together with production management includes group discussion sessions on such subjects as the female work force. The foremen also meet in two-day sessions in groups of fifteen to look at the future role of the foreman. All members of works councils and joint consultation groups are given a one-day training course in conference and consultation techniques.

A new workshop opened recently at Köping, giving employees a chance for more innovation. The building is new and so are most of the sixty workers. (Fifty of the new employees are women.) The workshop has a new manager, who is encouraging workers to choose their own form of organization. There is no foreman in the workshop; instead, the employees elect their own leaders. The project was planned by an informal task force that included experienced operators from other parts of the plant. They don't include any follow-up or evaluation of the self-managing workshop in their plans, on the premise that a more casual atmosphere will be more conducive to trial-and-error development. This attitude permeates other changes at Köping. Very little of the change process there is enshrined in plans, memos, reports, or formal studies. Face-to-face communication is considered more important.

The Arvika foundry, built in 1970, produces about 25,000 tons of castings each year. It was advanced for its time, with as much control as possible for the noise, soot, dust and heat that bedevil all foundries. Although the designers believed they had done everything possible, an employee-rooted renovation project in 1972 demonstrated further improvements. Adapting an idea that had been used at the Skövde foundry, for instance, the project group installed new "cleaning cabins" for sandblasting, chiseling, and polishing castings from the foundry. Special lighting, ventilation, and soundproofing measures make the cabins themselves pleasanter to work in than the old open stations, and the noise levels outside the cabins dropped dramatically. Workers can work alone or in pairs in the cabins, as they choose. Each cabin costs about $11,000; productivity in the department increased by 15 percent after they were installed. In another area, mechanizing a noisy cutting operation improved the noise and dust conditions and resulted in a 60 percent increase in productivity.

The most difficult jobs are almost always the most expensive to mechanize. Arvika is now working on a way to trim and clean castings in a trimming press process, eliminating the traditional task entirely. This has already been achieved for differential housings and several other products. It cost about $170,000 for new equipment for three products, but manufacturing costs dropped to about 40 percent of their previous levels. Trimming press tools for other products are planned, which will result in further reduction in dust and noise problems. Workers are involved in deciding which operations to mechanize first.

## HÄLLBY AND ARENDAL: NO ASSEMBLY LINES

A new plant is a new opportunity, as we learned at Kalmar. The Hällby tractor factory at Eskilstuna, on the shores of Lake Mälaren outside Stockholm, was inaugurated in 1975. By this time we had carrier and buffer concepts firmly in mind for any new facility. The warehouse at Hällby, for example, is highly automated, with a carrier-type device that climbs between high stacks of parts, eliminating the reaching and lifting that people had to do in earlier warehouses.

In the assembly area, teams put together a complete tractor on a carrier especially adapted for welding and turning or tilting the cabs. This operation has no conveyor line at all. The platform stands still and employees can move it as they wish. The same concept is used there to make forestry machinery.

The Arendal truck workshop, near Torslanda, is one of the most interesting from a human viewpoint. We are able to make changes inside the confines of an existing building, and to try out new ideas for a future truck factory.

The assembly line has evaporated. Instead, a team of nineteen people assembles even the heaviest trucks, at the rate of about two a day, using a carrier based on the old air-cushion concept. Interfactory cooperation has been an important feature of both these factories as they developed. Arendal engineers and supervisors started thinking about changes in 1974, and rapidly involved employees in the project. Usually when engineers are designing a new facility, they create new and more sophisticated tools. In this case, the workers told the engineers what tools they needed—and these turned out to be less sophisticated, not more so. The result has been extremely high quality—with less expensive tools. Furthermore, the workers take care of their own tools, so tool maintenance costs have dropped visibly. Before the Aren-

*Air cushion carriers.*

dal project started some of the engineers and workers visited the Lundby plant.

The Hällby engineers worked together with their Arendal counterparts, and from the beginning set up a reference group including members of the work force. They adapted the air-cushion carrier idea for their one-team, one-tractor production. It is quite normal at these plants to see two men effortlessly pushing together two nine-ton elements to create a single eighteen-ton truck.

The benefits of these investments are reduced production cost and higher quality products. Furthermore, we gained another indication that team working was welcome to the workers. Absenteeism at Arendal dropped dramatically to about 5 or 6 percent.

## DOES IT WORK OUTSIDE SWEDEN?

Volvo management realizes that the easy acceptance of joint union/company projects is uniquely Swedish. The company's emphasis on improving work content was also quite early in terms of the current worldwide trend. So, when we bought a controlling interest in the Dutch auto manufacturer DAF in 1975, we wondered how many of our ideas would transplant themselves. As it turned out, the Dutch at DAF were suffering some of the same problems we had observed at the beginning of the seventies and were thinking about some of the same solutions, although their approach was more "experimental" than ours. The association with Volvo simply added impetus to a trend that was already there.

Volvo Car BV, as the Dutch operation is now called, had its factory at Born in the depressed Dutch mining area near the German border at Limburg. Volvo already had a plant at Ghent, an old Belgian city rather near the Dutch border.

The Born factory, which now manufactures our smaller model 343 car, employs about 3700 people. Slightly over a third are foreigners because there has been an outflow of local labor to Germany where wages are higher. First Belgians were recruited for Born, and later people from Morocco and Tunisia.

*Air cushion in Hällby truck assembly.*

The Born plant has one new feature that is unique in Volvo —two chiefs (or managing directors), one from Holland and one from Sweden, to facilitate mutual understanding. Production was almost 100,000 cars in 1973, but dropped after the oil crisis. At the same time, absenteeism and employee turnover rose sharply, and managers began to think about new forms of working. A steering committee of management and technical people set themselves the task of improving work on the assembly line by the end of 1973. To oversee the experiment, the steering committee set up a nine-person task force, including a representative of the works council.

The task force chose a relatively difficult portion of the Born assembly line and with the twelve workers there explored job enlargement. Although the work group was mixed in sex, age, and nationality, involvement grew rapidly as members taught each other their jobs.

The task force noted increasing and spontaneous offers by workers to help each other. By the end of 1973, most people had learned from three to eleven of the twelve jobs and could work in longer cycles, thus creating longer rest periods. The foreman, together with the group, took on planning the work—a job formerly done by the foreman's supervisor.

While most of 1974 was officially spent reviewing and evaluating the results of the first experiment, similar activities started spontaneously in other parts of the plant, and these were not discouraged. In early 1975, a new steering committee was formed, including not only managers of all line and staff departments but also representatives from the works councils. They set up a structure of discussion groups involving a large portion of the work force. Today about 150 groups of fifteen or so meet regularly, about once every five weeks; more than 80 percent of the blue-collar workers take part in the meetings. Management development courses are also planned, starting at the foreman level, concentrating on interpersonal relations and communication.

One outgrowth of the first Born experiment is a new assembly hall, where the conveyor belt is divided into small parts, with intermediate buffers between them. Workers control the overhead conveyor when they want something to move. The idea is now being studied for Torslanda. Automation and buffer stores at Born have increased flexibility in the body shop. Management has announced that there will be close cooperation and early information to employees whenever new equipment or layouts are considered, and a "letter of intent" has been sent to all employees and discussed in all work groups, setting out management's desire to improve the work structure not just as a project but as an ongoing process. Born management now contends that the experimental phase is over and the factory has found useful development strategy for the future.

Less troubled by absenteeism, employee turnover, or foreign workers, Volvo's own Ghent factory was somewhat later than Born to begin reexamining work structure, but managers are pleased by recent progress.

In April 1974 Ghent management decided there was a need to make the individual more central to the production process. Four factors influenced this decision:

- *rapid progress in most of Volvo's Swedish plants,*
- *a new emphasis on the quality of working life in Belgian unions and media,*
- *a strong and growing demand from workers for better conditions,*
- *the need to improve the quality of products.*

The first phase of change, mainly instigated by management, focused on better arrangement of relief people and installation of buffer areas. Instead of one relief person for twenty-five workers, Ghent decided to have one for every fourteen workers. Where the system's designers could use buffers instead of relief workers, they aimed for storage that would provide twenty-minute production coverage. Pits in the assembly factory and merry-go-rounds in the body assembly plant were replaced by fixed stations and group working. Success with job rotation in the upholstery area led to abandoning the line entirely there in favor of group working. Today a worker assembles an entire seat and takes responsibility for the quality.

In the second phase, automation and group working trends continued, and the managers began to realize that those areas where workers had had some input to the change process were obviously running more smoothly than those which had been changed entirely by managers and engineers. Now top managers are looking for better ways to keep in touch with workers. Because participation demands stronger, more confident leadership, management education has a higher priority at Ghent today. A "group working" project team, including workers as well as foremen and engineers, will investigate further changes. Ghent management says the results of the program so far are convincing.

These are not the only steps toward participation in Volvo factories. They are simply examples that have been documented to some extent or have come to our attention at headquarters for various reasons.

It is heartening to discover new examples every time we go out to visit a plant. To me this indicates that our central commitment has transmitted itself to the farthest reaches of the company. The momentum first hoped for is beginning to be evident, even in factories outside of Sweden. At the same time, Volvo has avoided the trap of a corporate-wide "program" that might limit the imagination and commitment of people at their own sites.

The management style has changed not only in Volvo's factories but also in the offices. One of the most important changes was to streamline headquarters from 1800 to about 100 employees. Former headquarters people from such specialties as personnel or public relations say their own jobs are more rewarding when they are closer to the problems and can more easily see the results of their work in local terms. At the same time, they and their know-how are more welcome as members of the local unit rather than emissaries from headquarters.

## RULES OF THUMB

From where I sit, it becomes clearer every day that the most effective changes are those in which the workers themselves had the largest hand. It is almost alarming to realize how much know-how and capability has been locked up in the work force, unavailable to managers who simply didn't realize what an important resource it was.

Our experiences at Volvo have changed my view of management somewhat. Unlocking worker potential has become as important as any display of brilliance in technical terms. Management development becomes a critical factor, then, in keeping up our productivity and making our factories and offices better places to work.

Volvo has no "management development" chief. The task is too important to be put out to pasture in a specialist department where it can be ignored. Instead, it should be viewed as one of line management's most important tasks. This view is increasingly shared by managers throughout the organization.

Their foremen, supervisors, managers, and workers are resources for which they are accountable, just as they are accountable for investments in buildings and machines.

In this atmosphere of employee participation and rapid change, management is an exacting task. If you don't manage tautly you can drift into inefficient "permissiveness" that endangers the entire venture. Instead, it is the manager's job to create an atmosphere, a sense of urgency. Tight management need not be authoritative. Today's manager must be able to talk to people, and to listen as well. If the manager is alert to every opportunity for improvement, and full of zest for the job, this communicates itself to others. Participation demands more work, not less, from everybody. Idle people become bored and sloppy, so it is an important part of the manager's job to be keeping tabs on group climate all the time, injecting some of his or her own alertness whenever signs of apathy or boredom arise. As Volvo and other companies have learned, the manager who is reluctant or just giving lip service to the idea of participation can hold back employee-based changes that are actually in the best interests of the corporation as well as its workers.

The changes in our various plants give rise to a few rules of thumb that may be helpful to others.

- *Each unit should be free to develop individually, without excessive control or interference from headquarters.*

- *An active and positive top management involvement in change is a prerequisite for positive results.*

- *Headquarters is most effective if its role is to sanction investments for new approaches and to challenge local managers to take more radical initiatives and risks.*

- *Positive achievements seem related to the extent that managers understand that the change process sooner or later affects several organizational levels, regardless of where it started.*

- *Problems crop up if we attempt to formalize change, requesting targets, minutes, and figures too early. Change requires time, and freedom of action. When people view it as a continuing search and learning process of their own, the chances of lasting effects are increased.*

- *The initiative for change should be a line responsibility, with specialists as supporters rather than initiators. Changes they initiate seldom have lasting effects. They can, however, act as sounding boards and catalysts, carrying know-how from one place to another.*

- *Progress seems to be fastest when a site or company starts by forming a joint management/union steering committee to look at its own problems.*

- *Steering committee members should be the strongest possible people, with shared commitment to change.*

- *The fastest way to get ideas flowing seems to be setting up discussion groups in each working area. A working area in this sense, and in a group-working sense, should probably contain fewer than twenty-five people.*

- *Groups that have money to spend on their own facilities and a mandate to list their own problems seem to achieve cohesion and cooperation more rapidly. This does not need to cost the corporation more money because facilities or safety budgets can often be apportioned to the groups themselves instead of to specialist departments.*

- *A new plant, a new product, or a new machine is an opportunity to think about new working patterns.*

- *An investment in one new facility or one group area often results in spontaneous changes in related facilities or groups. These can be encouraged by alert managers.*

- *Most factories have a number of tasks that need not be done on assembly lines. Once a few have been found and changed, others will reveal themselves.*

- *Changes of work organization must be integrated with a structure of employee consultation, so the change suggestions emerge from inside.*

Some of the most effective changes in work organization at Volvo have taken place naturally, without projects, without scientific sophistication, without being reported to anybody. Those changes occur simply because people are keen and interested. Finding ways to encourage such changes is management's challenge.

# 6

# The businessman and the public

**A**t every point of contact between business and government, there seem to be differences in values, attitudes, and assumptions. As a result, it becomes difficult for the two most important institutions in society to carry on useful dialogue. And dialogue with the public itself is hampered too.

Most business people are convinced they are doing the right things for the right motives, and would like to do so with a minimum of government interference. The press gives the impression that most people in government seem equally convinced that more controls must be imposed to prevent business from abusing its power over consumers, employees, and society. The public, bombarded by too much opinion and too little real information, tends to support the antibusiness attitude.

## PUBLIC MISUNDERSTANDING

The problem for business people has two aspects. First, the sense of being misunderstood engenders defensive attitudes, retreat into hurt silence; thus, unfair accusations as well as those that are fair are both rejected in most cases. Public criticism helps business hide from the real question: Are the values of business up-to-date? I contend that there is a need to reexamine business values to keep up with changes in public attitudes.

The lack of self-examination in business simply adds fuel to public mistrust and makes real communication even more difficult. Even though a number of businessmen are aware of the changing balance of stakeholders, the public still seems to believe business is interested only in profit for the shareholders. This fierce image further damages what chances exist for communication—the second aspect of the basic problem. Until business can shed its defensive shell and begin visibly changing its own assumptions, it will have to go on living with adverse public reactions, which in turn make it difficult for business to be introspective.

It is not difficult to see how antibusiness attitudes arise. As our social structures become increasingly complex and urban, control is exercised by larger and larger institutions, in both government and business. It becomes almost impossible to find a gas station that cleans your windshield, or a grocery store where clerks recognize you and smile at the checkout counter. People move more often, which creates a demand for the comfort of identical shop fronts in every suburb, the trademarks of large companies. Consumer issues focus on the negative aspects of business. And the response from the business community is usually ostrichlike in its eloquence. There are steps the business community can take to reverse these trends and to achieve more healthy dialogue with the public we are really chartered to serve. These steps include more open information, creating smaller organizations, finding means to achieve accountability to the public as well as shareholders, and developing clear-cut personal and corporate sets of ethics that demonstrably govern our actions.

## OPEN INFORMATION

Public misunderstanding is not uncommon when the press and politicians take up what are essentially technical or business issues. In both Sweden and California, for example, the 1976 elections were permeated with the question of nuclear power stations. In the heat of campaigning, the issue grew emotive. Politicians made charged statements. Newspapers had to simplify complicated matters with large headlines and small articles. What facts there were often remained buried in obscure technical reports.

What the public really needs for proper evaluation of such issues is clear, objective summaries of the technical information available, interpreted in unemotional terms, telling them what the results of various approaches are likely to cost, what their effects might be, and the probabilities associated with each. Such information needs to be dispersed as widely as the political statements.

The situation is not the fault of television, newspapers, or magazines. They have limited space, limited manpower, limited time, and limited budgets. They do as well as they can given these financial and social constraints. Most business people would occasionally disagree with this view, but that, too, would be scapegoat-hunting rather than fact-finding. A more fruitful approach might be to analyze the information the media need, and then help provide it as honestly and fairly as possible.

This is often easier to do regarding an outside issue than regarding a question that affects the company. The way business works is often hidden from the public, sometimes because of organizational reaction to past press criticism. So it is natural for reporters and editors (and thus readers) to assume that business has something to hide. Furthermore, because business issues are complex and often rather boring when experts present them in their entirety, reporters find it easier to get quotable interpretations from workers, housewives, or politicians. They know less about the issue, but they talk in a more interesting way.

Like companies, government departments often cloak their internal activities in secrecy. They also suffer the same inability to reexamine their values. Why don't they get blamed as often as business for social problems? For one thing, people tend to think of "the government" as a single unit, and that unit has many vocal outlets among its elected politicians. Their commentary often masks the silence of government departments. In addition, the government clearly belongs to the nation, while big business is often involved internationally.

It will take training to change business attitudes towards outside communication. Unlike money, information is not lost when it is given to someone else. Yet the culture of management has evolved in an environment where distorted information sometimes caused loss. This is one basis for closed atttitudes in business towards open communication.

Another reason managers withhold information from people, particularly employees, is that the "owners" of the information feel the employees would not understand. This elitist view may have been valid when factory and office workers were undereducated and underprivileged. Today, however, this is no longer the case. This lingering parent/child attitude infuriates the adults who make up most of the work force (and the public), and I believe it underlies some of the conflict in modern industry.

Instead of training managers to treat information as status-conferring "property," to be protected and held close to its owners, companies would benefit enormously if they could retrain managers to measure themselves in terms of their skill at communicating and interpreting to employees the mass of information that needs to flow inside any organization. This role is difficult to achieve because it requires managers to unlearn attitudes that are deeply ingrained in the behavior of most organizations. The re-learning must be bolstered by example much more than by words. The chief executive must set the example for candor long before one can expect to see it at other levels.

This management barrier is probably the single most important hindrance to openness, and thus to effective communication inside and outside the company. If it can be overcome, powerful changes are possible. At Kalmar, for example, the foremen are no longer "bosses" but instead view themselves as resources, men with special skills and special training to answer questions, help people learn new jobs, help the workers learn to make decisions for themselves. It has been a difficult change for the foremen, but their commitment to the changes evidences the increased personal satisfaction they derive from this role.

Although the process of change is difficult, it is possible to establish such a strategy inside an organization and develop new attitudes towards information. Top management has less "control" over what happens outside, but that control is already illusory. If reasonable progress is possible inside the organization, why not try outside too? Are politicians any less intelligent than factory workers? Is the press incapable of understanding a single company, or business in general? The changes in attitude to outside information will be just as difficult as they are for inside information, and they will take a long time, but useful dialogue is a goal that justifies the effort.

## BIG IS NOT BEAUTIFUL

For several decades, the trend toward super-scale organizations has been nurtured and justified by economic and technical factors, not human factors. As our organizations have evolved—like our production lines—into huge, inflexible structures focused completely on technical and economic goals, their weight alone works against change. Big organizations simply do not welcome change; it disrupts their stability. Nor do they foster new ideas about management, because change would influence the working routines of most members of the hierarchy. And the cost of changing attitudes is discomfort. Therefore, changes are opposed or rejected too easily.

In today's climate, top executives in large companies are aware that this resistance to change is hampering the corporation's ability to gather valid information about the outside world. They often want to decentralize and delegate more responsibility to employees. Yet they constantly meet this built-in obstacle, this wall of reasons why proposed changes won't work.

Most top executives feel such problems are unique to their own organizations. Partly because business has become so competent at hiding its failures and problems, the public (and even some business people) are unaware how common the problems of hierarchies have become.

Trapped in mazes of systems that once were adequate, business responds to problems in one system by creating another system that brings even more problems. While such technical systems freeze practice, the power and responsibility inside a living organism like a company are constantly shifting. One can see the same effect in the world's currency system, a highly technical system policed mainly by the United States. But today it is out of step with the real power bases in the world's money. In economic systems or corporate systems or social systems, the more diluted the power and the more technical the system, the more problems it seems to suffer. This fact does not contradict the benefits of real decentralization, because a properly decentralized organization has clear-cut boundaries, so each element has the power it needs to deal with its own responsibilities. The problem of the over-technical, systematized solutions is that they cannot move to keep up with changes in the environment when a disparity grows between the de facto power structure and the one for which the system was designed.

Smaller enterprises, by their very nature, usually give their employees more freedom in what they do and how they do it. With less automation, such firms tend to be more dependent on their workers. In a small company an employee can do something new without having someone in authority immediately call him or her to order. Some of the practices of the small company could help big companies. This is what Volvo is trying to do in its new factories. To limit size without giving up the real economies of scale, we look for ways to create smaller workshops inside a single factory.

Our experience at Volvo supports my belief that we have plenty of room for individual choice inside our industrial structure. The individual is capable of planning and executing his or her own work. A worker can indeed judge personal performance, measured in its own terms, not related to some statistical average.

Where production is concerned, business has taken it for granted that working discipline is necessary and that yardsticks are needed to measure achievement. We measure output, develop standards, and pay wages somewhat in proportion to the work that is done. These yardsticks are fairly easy to develop in an assembly-line factory. One can count the number of screws turned, fenders painted, or cars produced. Group working, though, means developing group yardsticks.

The task of developing yardsticks becomes much more difficult when you walk out of the factory and into the office. More and more work today is administrative. The more we automate blue-collar processes, the more visible becomes the shift toward white-collar jobs. This means there is much more work for which we have no valid yardstick.

We talk a lot about blue-collar work and its problems, but white-collar work is perhaps even more boring and dull today. Manufacturing jobs are at least basic, with a visible relationship between what you do and what comes out on the other end. But there is no such relationship if you are a punch card operator or shuffling papers in a social services office. A great deal of restructuring is called for in white-collar jobs, for young people are beginning to reject these kinds of work more and more. There is no one who wants to wash the dirty dishes in a school or restaurant or home.

The administrator who does not feel well-regarded for his work may need to invent yet more forms of reassurance. The person who surrounds himself with bureaucratic processes that are cumbersome, but measurable, is trying to create his own yardstick. He is not a villain but a victim.

Throughout the sixties, efforts were made to automate the office the way Henry Ford started automating the factory. Technicians developed machines as fast as management could conceive of them. Then, supervision and control mechanisms had to be created to keep humans from making mistakes with the machines. In the paper factory as in the production factory, the only role left to the employee was to serve the machines—a very negative view of the value of a human being.

Today, spurred to some extent by social unrest, we are questioning assembly-line concepts in both the office and the factory. We value the human more highly in the context of working life now, but it will take time to change from the large units and systematized working patterns of the sixties. Most people today, at Volvo and in most other organizations, have to work within some kind of system that sets restrictions. Even if they could be freed from unnecessary measurement, many workers would want some external symbols of their achievement. So, we need to find personalized ways to set individual performance standards, in the context of clear-cut, useful assignments, inside smaller-scale organizations.

When we analyzed employee turnover within Volvo, a clear correlation between high mobility and large organizational units became apparent. We already realized this was true of cities versus smaller towns—rates are higher in Stockholm or Los Angeles than they are in Umeå or Wichita—but we learned from this analysis that the same rule of thumb held true for working units themselves. The larger the factory or office, the higher the employee turnover. This was in complete contradiction to the technical viewpoint that has ruled our lives for so long. In truth, economies of scale are limited by human factors.

We are beginning to realize that a company depends on people, not machines. People, in turn, depend on work. But they want work that is human, in working units that are smaller.

## THE SOCIAL AUDIT

The role of business in society merits clear definition. Otherwise, we may find it defined for us by outside factors. Business needs to face up to this question, like many others, with openness and increased dialogue.

A company interacts not just with its shareholders and employees, but also with governments in every country where it does business, with citizens in every community in which it maintains offices or factories, and with the consumers who buy its products. We also interact with other colleagues in business, so our own values influence a great many people. It is important not only to be aware of these interactions but also to find ways to evaluate them and keep track of them, to make sure that in all contacts with the public we are contributing to its well-being. Only then can we reasonably expect it to contribute to ours. And when we know our own relationships are healthy, we are likely to have more energy to spend on issues that stretch beyond the limits of the corporation.

One means for gathering information about our own relationships as well as our interactions with the outside world is the so-called "social audit." In some instances it becomes an elaborate quantifying exercise, perhaps consuming more energy than it channels. Sometimes it is more a social justification than a social audit, too. But properly done, it can be useful. In a number of American firms the social auditing function takes the form of an ongoing group, usually at board level, sometimes called the corporate responsibility committee. However, at this point I am not so much espousing the doing of good works as the collecting of good information. What *are* the social problems of the community, or the nation, and what is the company already doing about them? Only when these questions have clear answers does a firm have a real basis for long-term planning that takes the outside world into account. Similarly, if a company has a clear view of its own problems, it has a chance to do something creative about them, as Volvo discovered when it began to plan the Kalmar factory.

Taking the outside into account is not always comfortable. I recall one company with which I was concerned which was able to plan a new kind of community because of this kind of social awareness, but failed to build it because the board did not share the management's values regarding the outside. When the government required firms in its industry to invest in housing bonds, the company decided to take an active interest and see what it could do to help alleviate the problems of alienation, mobility, and lack of common interests that characterized so many new bedroom communities. Instead of merely lending money, the company asked for full responsibility for a tract of land that would house 20,000 to 30,000 people, the optimum number its research showed could support all the necessary community services. The company hired an architect who designed a town that would integrate working and living, with appropriate living quarters and part-time job banks for young people, pensioners, families with young children, and so on. Thus, most of the work in the community could be handled by its own members. The scheme was ultimately killed for in-company political reasons, even though the same investment had to be made whether the firm's participation was active or passive.

To take action on a wide front like this requires heightened awareness of public problems, particularly at board level. To get better information at board level in Volvo, we have widened the membership of our board. In many companies the board of directors is still a group of people who have known each other for many years, with shared values and assumptions that are seldom questioned. In 1971, several years before worker-directors were required by law in Sweden, Volvo invited first two, then four people from the shop floor and offices onto the main board. The most important benefit was not necessarily for the employees "represented" on the board so much as for the existing board members who began to question their own assumptions and present their views more clearly. Then, in 1976, Volvo brought onto the board a representative of the community in Gothenburg, our headquarters town. He should add new insight for all the other directors, as well as bringing information about our community and the public interest that would have been difficult to obtain otherwise.

Similar phenomena—consultations that bridge organizational gaps and company walls—give similar benefits at other levels of the corporation. A social audit, or ongoing process of gathering and evaluating information about the company's activities and impact outside, as well as about the community's needs, can sometimes help such information travel faster.

# THE BUSINESSMAN/CITIZEN

Business exists to serve the public. This basic assumption is sometimes obscured even from business people themselves by the difficulties of day-to-day operation and the defensive attitudes they develop when business is criticized. To operate effectively according to this precept, the modern corporation's basic strategy should be based on an outlook that permits it to act aggressively, taking the initiative with the conviction that the corporation is serving a just cause. This is why the enterprise needs its own set of values that are honestly established and openly declared.

The individual within the corporation also needs a set of values that are visible and workable. Each person in business, as an individual, has the right and the responsibility to work out what he or she believes is the right approach to public questions, whether they be political or social.

Within the political realm alone, business people may have quite a bit to offer the public. It is possible to bring to political questions some of the tools of business management. I have personally found some of the perspectives of business to be helpful when looking at politics.

In business, for example, managers must learn to manage large projects and master the skills of administration. They quickly learn to sort out long-term and short-term objectives, too. In general business practice, the strategic questions are those dealt with by the board, and the tactical questions are the fabric of day-to-day management.

Similar types of questions arise in politics, although we have no institutions beyond an informed electorate to deal with the long-term questions. This means that politicians must often make decisions based on short-term perspectives, but the consequences of their decisions are going to be much longer lasting. So many small decisions clutter up people's lives with new forms and committees, or transfer finite bits of power from local government to regional or even national levels that it sometimes seems that the net result of government has been to erode individual freedom of action. It is important to safeguard the liberty of the individual as much as possible, in work and in every other aspect of life. Too often action justified on the grounds of "public interest" is carried through at the expense of private autonomy.

Politicians are not to blame for this situation. They are, after all, acting according to the systems we have set up for them. It is unfortunate that they have no long-range, strategic "board" to which they could refer the questions that obviously have long-term consequences.

The education system is one example of the disparity between strategy and tactics. It can take up to ten years between the time an educational reform movement starts and the time the reform is implemented. Educational reform is a fundamental issue that affects at least several generations. Thus, education policy should not be swayed by short-term political and tactical considerations. Nonetheless, it happens. The administration of the education system, on the other hand, is a different matter—one of "day-to-day management." Once the goals and guidelines have been laid down by the voters, administrators must make the best possible use of available resources to meet them.

National housing and defense needs suffer problems similar to those of education. The effects of political decisions transcend terms of office. And, as in business, those who created the system or the reforms are not necessarily those best qualified to administer them. However, in the public sector these two tasks too often merge in a single governing and executive body, in fact if not in name.

Decisions taken by the current government will influence events long after the current terms of office are over. Therefore, we need to find mechanisms for putting clearly defined goals and long-term action programs before the voters, to make our elections more meaningful. Perhaps modern governments should, like modern business, expect their structure to change as the world changes. Governments might look for ways to have two-tier cabinets, for example—one concerned mainly with administering the country's resources as efficiently as possible within the framework of long-range political objectives drawn up by the other tier. Those objectives could become a party's election platform.

The public and private sectors often require different kinds of technical expertise and, in many cases, their requirements and goals are visibly different, so a sharp line between the two seems natural. At the same time, though, such a demarcation should not imply that the two sectors cannot use each other's experience. Prejudices sometimes keep this from happening as much as it could. The private sector nurtures a long-standing suspicion that public administration is inefficient and operates in idyllic circumstances free from pressure. One of the many corresponding illusions entertained in the public sector is that the private sector in its drive for efficiency invariably gives economic objectives priority over human objectives. Both views seem to stem from a lack of contact between the two. In my opinion, both views are excessively categoric, and consequently mistaken.

The gains to both sides from cooperation are considerable. Within a corporation, people participating in public-sector activities may feel they are making a worthwhile contribution—which in pure management terms usually makes them more motivated in their other activities as well. The public recipients are less likely to express appreciation than the private donors when this cooperation takes the form of "good works." A two-sided partnership is more likely to be fruitful in the long run.

Often, the first visible result of community/company cooperation will be an increase in the number of complaints flowing in from outside—just as Volvo experienced with the initial steps toward employee participation. However, it is easier to view this as a healthy side effect of cooperation, or even as useful new information, when one realizes that the complaints have been there all along, festering without any means of expression.

Over the long term, from a company's viewpoint, partnership with the public is simply a good investment, because there is some chance that society and thus business will function better for it. To rephrase the often-misquoted Alfred Sloan comment, most larger companies these days are beginning to realize that what's good for America is good for General Motors.

In a number of American cities, private interests have set up consulting or survey groups that the city can use to increase the effectiveness of its administration. Britain has several clearinghouse organizations, privately funded, to match the needs of public-sector or volunteer agencies to private-sector donors. The experience is usually stimulating as well as educational for volunteers. Society gains by better public administration. In New York, for example, people from the private sector helped draft new working procedures for the courts that drastically cut the time between arraignment and trial. The immediate benefits are obvious, but the most important result of this kind of collaboration, in my view, is better mutual understanding and confidence between the public and private sectors.

It should be possible for business to respond to the public need in its internal workings, too, to be steered within the framework of the community's social, political, and cultural values. But this requires the values of the electorate to be expressed in the mechanisms that govern production. These are still too often responding to short-term perturbations rather than long-term objectives.

The alternative to business working in harmony within a politically determined framework is socialization and government directives within a totally planned economy. This would not only erode the efficiency criterion we presently exercise in the use of our resources, it would also open the door to concentration and arbitrary exercise of power, stifling democracy by a compulsory reduction of available alternatives. I firmly believe that having rival alternatives is our guarantee of a living, working democracy.

# 7

# The future of work

I f things go on as they are—and I see little reason to hope otherwise—we are going to be coping with major social problems in the foreseeable future, problems that we can be preparing for today. I can speak with some experience about the problems related to one field, transportation. Most of them are already with us, at least in microcosm, and the situation in other realms is similar.

## TOO MUCH TRANSPORTATION

Transportation won't be the same in the future as it is today. Neither people nor local governments nor the auto industry itself are geared for the urban explosion that has turned many cities into infinite lines of cars, creeping more slowly than the horse-and-buggy transportation of fifty years ago. Automobiles consume expensive fuel, waste it sitting in traffic jams, and leave a murk of exhaust fumes in the atmosphere to mark their frustration.

The main culprit is, of course, the private car, which has not been subjected to any form of planning restraint. Nor do I think it will be restrained in the future, because a car for most people represents precisely the better standard of living they have been working so hard to achieve. So curtailing use of the private car will be impractical if not politically impossible outside the festering city centers, where even drivers are beginning to accept the fact that the auto is a nuisance.

Yet the car itself is not to blame. Outside rush hour in the city, the auto is the most convenient way to get from A to B. The contribution of the auto industry to our economic growth has been a major factor in twentieth century progress, too. It generates investment, is a major source of employment, and creates a demand for roads, bridges, and other investment and job-generating opportunities that contribute still more to growth. Most goods travel by truck and most people travel by car. Our industrial growth depends increasingly on our ability to transport ourselves and our goods effectively.

What are the alternatives to today's urban congestion? Although a few cities have poured vast sums into automated rail systems, these usually prove to be inflexible, expensive, and not always socially beneficial. Other, more practical choices, are likely in most cities in the future.

- *Upgrading conventional public transport systems for passenger comfort.*

- *Building more and better public transport terminals, with good interconnections, as well as ample parking at suburban terminals.*

- *Installing information systems so passengers can easily find out what alternatives they have to get from A to B.*

- *Subsidizing public transportation fares, possibly by levying a tax or toll on private autos using city center roads.*

- *Restricting the use of private cars in cities, at least during working hours.*

- *Improving public transport service to suburban industrial parks.*

- *Only marginal investment to gradually upgrade existing streets.*

- *Reserving parts of the city street network for rush-hour bus lanes.*

- *Designing traffic systems to make better use of vehicles presently on the roads, such as taxis, minibuses, and ordinary buses.*

All of these are within our means today, financially if not organizationally. In most cities, though, the department responsible for roads and traffic control is quite separate from the department responsible for public transportation, and the department that plans land use and factory location is yet another element, cut off from the road and public-transport authorities.

In transportation, as in so many other realms, I hope we can retain personal choice. However, I believe some controls will be absolutely necessary to save the city centers from being strangled by cars. The solutions don't need to be revolutionary. They can be achieved by using transportation and control techniques already available.

I don't believe we can or should "do away with" private cars. People have worked hard to earn their cars. More families can now afford two cars, and second cars tend to be smaller and more efficient. Having a range of choices is part of our democratic purpose, part of what we work for. Thus, we need to increase choices by providing a range of public transportation vehicles—buses, taxis, minibuses—so people will eventually begin to use each one for the things it can do best. This can only happen when public transportation actually offers better service than the private car. We will need to subsidize the public systems and control congestion during the transition until the public systems are sufficiently developed and accepted to compete on their own merits as a faster and better way of getting around a city.

The worst enemy of the private car is its misuse. It is unsurpassed for flexible, convenient, quick transportation between points—when roads are uncrowded. But when people use it for things like rush-hour commuting, for which it was not designed, it is a failure, and as a result the credibility of other rubber-tire vehicles like buses and minibuses suffers, too. The low priority most cities give to public transportation today has been a powerful self-fulfilling prophecy, forcing more and more once-competent systems to run down for lack of maintenance and investment, until they are less attractive and thus used less, and thus still more impoverished, so that more people decide to use their own cars, no matter how long it takes.

Volvo has been involved in the design of an experimental taxi which was exhibited in the summer of 1976 at the Museum of Modern Art in New York. Starting in 1972 with clean drawing boards and the idea that we needed a safe and practical vehicle for big-city use, the designers came up with a useful compact vehicle for passengers and drivers alike. The design can be adapted for disabled passengers or minibus service, among other things. It includes an ergonomic driver's compartment completely cut off from passengers, with excellent visibility.

*Volvo's experimental taxi.*

The taxi's turning radius is less than thirty-four feet, achieved by mounting the wheels "one at each corner." The passenger compartment has a sliding door for easy entry, low door sills so wheel chairs can be put in easily, and a safety bar rather than seat belts.

In transportation, as in so many other realms, the main constraint to changes like these is not the technology—we can already demonstrate most of the advances that will be necessary. Instead, the constraint is likely to be individual and organizational inertia, suspicion of the unfamiliar, and apathy. This concerns me, not only because my own work is in transportation, but because transportation will remain an important part of most people's working lives until we can reach a more ideal integration of our home and working lives. We must try to cut the overhead the individual pays—on his or her own time—in getting to and from work. This is a social cost that never enters into official calculation, but its importance is growing. The average person these days budgets nearly an hour, each way, for getting to and from work—a total of nearly ten hours a week that could be spent either in fruitful work or pleasurable leisure. Flexible working hours might iron out peaks, but they work somewhat against car pools. The best long-term solution will be to pay much more attention to integrating home and work environments. This means having smaller factories and more of them, near where people want to live, rather than putting huge soulless housing developments where the factories want to cluster. There will still be individuals who choose to live in one community and work in another, but this is a personal choice that carries its own cost in commuting time. So far, choice is seldom available, so the individual and the car bear the burden for poor public planning.

## THE RIGHT TO HAVE A JOB

The utopian ideal of having home and work well integrated seems almost absurd in a world where unemployment—finding any work at all—is a major question for many people. Yet the ideal should be kept in mind, even while coping with the current problems.

I believe unemployment is not just a current problem but a long-term situation that is going to demand new kinds of thinking before it can be solved. Structural reasons explain some of the rise in unemployment in most industrialized countries, as do the short-term economic cycles that create employment peaks and valleys.

The employment question is going to be one of the major problems of our time. Quantitatively, we want the number of job opportunities in a country more or less consonant with the number of people who want to work. Achieving this goal of full employment will require qualitative changes in how we organize work, the conditions under which it is performed, and the attitudes we develop in ourselves and our children about it.

To a physicist, the term "work" means the conversion of energy. But work as a human being performs it is far more complex. To a person with a dreary, repetitive job, it is an ordeal that must be survived in order to earn leisure time. To someone with a consuming interest in a job, leisure is sometimes an unwelcome or rejected distraction. In days gone by, people did more physical work; now we have made advances in technology and harnessed new sources of energy to diminish the need to measure work in physical terms. Sometimes, though, we still measure the living standards of nations in terms of the quantity of energy consumed by each person in a year.

In today's world the strains of work are more often mental than physical. Although we still find occasional cases of people being worn out by hard work, it is more common to hear complaints about sensory or psychological circumstances than undue exertion.

The demand for physical labor is certainly diminishing, and some believe that work will become a privilege within the next few decades. The drop in demand for physical laborers marks advances on other fronts. A supertanker, for example, carries 200,000 tons of oil, yet requires only a skeletal crew aboard. In the clipper ship days, crews were fairly large. We ship perhaps 10,000 tons of cargo per employee today, compared to one ton per employee a few decades ago.

Or take another activity—making computers. One company in this business says it takes only one-third the labor force it did five years ago to make a comparable quantity of machines today. The computers themselves now cost less per unit and do much more work than they did five years ago.

While the labor demand is diminishing, the working population is rising. A major cause of this growth is women entering or reentering the work force, exercising their new legal right to equal pay with men. In addition, more pensioners want to continue some kind of useful work to keep their lives interesting.

Most industrialized countries have in recent years raised the number of required years of schooling. A greater proportion of the school population goes on to college or university, too. This helps siphon off some who would otherwise be entering the work force early. However, it also raises expectations for work that is adult, interesting, and paid in keeping with the individual's investment of time and money in education. The use of increased schooling to keep people out of the work force (which was never its purpose) was of limited duration, as we are beginning to learn now with the rise of unemployed university graduates in most industrialized countries.

Education is an invisible asset. Sweden spends more money on education, per capita, than any other country. The American investment is huge, too, and includes a large private-sector element. What is our return on this capital investment? At the moment, as new university graduates face unemployment even before they enter the work force, they must share the concern that we are not making good use of our investment. Today graduates must be prepared to take any job that is available—in essence, they are competing for jobs with high-school dropouts.

Social reform has had some unplanned effects on the demand for workers in industry. Most industrialized countries give decent support to those who are sick or out of work. In Sweden sick pay is 90 percent of an employee's normal pay. This gives every worker the decision regarding whether or not to go to work, at very little economic cost to the individual. One cost of this social reform is absenteeism.

For an 8000-employee facility like Volvo's Torslanda factory, a 15 percent average absenteeism rate means the company must employ about 1200 people more than would actually be needed if everyone came to work every day. The rate has doubled in the past ten years. If the rate doubled again, this would add to the cost of the average car some portion of the extra expense for employing 2400 people more than are technically "needed" in that factory. This is inefficient, inflationary, and it makes planning and managing production difficult. It is also "lumpy." Having enough people to cover the peaks means that many are underemployed on days when the absenteeism is lower. This situation penalizes those who come to work regularly by impoverishing their jobs to make work for the extra people between absenteeism peaks. At Kalmar, about 20 percent of the employees were responsible for about 80 percent of the absenteeism. The hardworking 80 percent deserve to have workloads that don't suffer day-to-day fluctuations.

Our solution to this problem might be the adult (but administratively difficult) approach in which each employee makes his or her own contract with the employer for a certain number

*Women are valued work-group members.*

of hours of work per day, perhaps in two-hour elements. Many contemporary social problems are related to the fact that people have few alternatives to the normal eight-hour working day. This costs the taxpayers money in sick pay, unemployment benefits, and so on. If alternatives to the eight-hour day were established, it might help alleviate a number of the problems and costs for the community as well as the company.

If it were possible to treat workers as adults, making their own contracts with employers to work a certain number of hours per day or days per week, it might be necessary to offer extra pay to those who chose to work the "normal" eight-hour day five days a week. Because the money for that kind of "bonus" would be coming from the public purse, in lieu of what are today sick pay or unemployment benefits, this is the kind of question that merits public debate.

Politicians are in charge of the policies that pay people almost as much for being sick as a company can pay them for coming to work. It is thus up to the politicians to decide whether to redistribute some of this money as incentive pay for those who choose to work hard under some kind of individual contract system. In any case, there seems to be a rising demand for some kind of work-hour flexibility from people whose family situations or commuting problems make the normal work day difficult.

Even in today's factories, of course, such a scheme would be difficult to administer. However, computers and meters for today's flexible working hours could easily be adapted to this purpose once people decided it was far preferable to maintain an adult employee-employer relationship, rather than casting one in the role of truant schoolchild and the other in the role of parent and taskmaster. Experience so far with flexible working hours seems to show that people come and go at quite regular hours each day, once they have the opportunity to select those hours for themselves within reasonable limits. When more companies have been able to base their production on work groups rather than assembly lines, more flexible forms of timing will be possible without the disruption that production suffers today from unplanned absenteeism.

Even if the job supply and demand are roughly equal, employment problems will continue. In Sweden we have more or less achieved a "full employment" situation, but imbalances still exist both in certain industries and in certain geographical areas. Some places have a shortage of labor, so people can choose their work. Other places and industries suffer a glut, so employers can choose their workers. Most countries have made sufficient social advances that those who are unemployed no longer starve, and there are only slight differentials in wages between the places with a shortage of labor and those with a glut.

The right-to-employment issue revolves around "the right to self-respect," rather than the right to avoid starvation. The demand for assured employment varies. In places like the United States, where average national unemployment figures have ranged from 4 to 9 percent since the war, and in portions of Sweden where there is a shortage of jobs or only sporadic demand, workers are increasingly demanding laws to protect jobs. But such rules often create "featherbedding" situations which are ultimately harmful to employment, as the United States railways demonstrated over the past several decades. With full employment people are naturally less interested in protecting specific jobs. Thus, the issues become somewhat politicized, with only those politicians representing high-unemployment constituencies paying sufficient attention to a problem that will eventually affect us all.

## THE RIGHT TO GOOD WORK

Employment is a problem that presently concerns everyone, and will be of even greater concern in the future. The advances that have brought about a real reduction in the demand for labor bring with them a need to reevaluate our attitudes to work.

We also need to change some of our social indicators. Practically every social system has developed economic models to measure what is achieved by the resources put into production. In most of these models, "production per employee" has been the criteria of achievement. But in view of the new problems of employment, one could begin to question whether these models will be useful in the future. If we continue to encourage employers to use fewer and fewer people per unit of production (and thus more and more machines), the effects on society could be serious. We may increase the suffering which unemployment brings to large groups of people, no matter how "humane" our unemployment benefits. American studies have shown, for example, that losing one's job has statistically significant effects on the lifespan and health not only of an employee but also of the spouse and family as well.* We need an additional dimension in our economic models, acknowledging that the other side of production-per-employee may be unemployment, which carries costs beyond the quantifiable benefits, in physical and psychological harm to individuals.

Simple price mechanisms of the supply-and-demand type only work to a limited extent. Furthermore, we find increasing competition between the performance of human beings and of machines. Too often the cost of human performance turns out to be higher than the cost of a machine's performance. This seems cruel—but it's true.

A few decades ago people had to work in order to stay alive. Even in affluent countries, the times of poverty, starvation, and misery are still within the remembrance of people alive today. An entire generation of Americans was scarred by the experiences of the Great Depression of the thirties.

Today we have achieved much higher general standards of living, and unemployment no longer means starvation. In fact, in some countries the difference between the useful income of a working man and of his neighbor collecting unemployment is negligible. So why work?

If you ask a worker in any factory whether he or she would like a holiday or not, the answer is a fairly predictable yes. On the other hand, if you ask that worker to choose between a six-month vacation or none at all, the answer becomes less predictable. Most people, whether they think about it or not, would probably have

---

* "That Helpless Feeling: The Dangers of Stress," by Douglas Colligan, *New York Magazine,* 14 July 1975.

trouble coping with six months off work. To be able to use the time off, a person might have to change his or her entire life-style. Many would find this difficult.

Similarly, if you ask whether people want to work or not, without economic penalty, I believe most of them would answer in favor of jobs. But having to choose might create confusion for some. Yet this is, in essence, what the social systems of most industrialized countries are doing today. It shows up in the attitude of young people, rejecting careers in anything, particularly in industry, before they even enter the work force. Coming to maturity without their parents' experience of the Depression, we find them travelling in Sweden on California unemployment checks, changing jobs for seemingly trivial reasons, or awarding themselves three-day weekends in the form of sick-leave, with alarming regularity. And who can blame them when we offer them petty, repetitive jobs without much human contact, often subservient to machines and systems?

If you combine these trends—towards reduced real demand for labor, narrower pay differentials, and higher unemployment benefits, along with higher educational standards—you find two of the most important issues for the industrial future:

- *how to create employment that suits people*
- *how to pay them as fairly as possible for doing it.*

The demand for meaningful work has become progressively louder, especially among younger people, for a number of reasons. Values concerning job content cross national frontiers more quickly than differences among national economies can be evened out. Most industrial nations are encountering similar demands for a more humane factory environment and more stimulating work. The demands for equalization in pay will probably follow the same lines, although perhaps more slowly.

In these circumstances, we can ultimately expect pay rates to become relatively equal in large areas of the world. Countries distinguished by their low pay scales today will become less attractive to international companies looking for the lowest labor

*Working teams need a range of age, background, and attitudes.*

costs. This will not necessarily be a disadvantage to industry. On the contrary, in human terms, the idea of locating a factory on the basis of local market requirements rather than a temporary wage anomaly would seem healthier in the long run. This could shift the bases for competition away from companies that can afford to move their production from one country to another and towards those that are simply most competent. Success ideally should be based on the ability to hatch new ideas, nurture them, and put them into practice competently, rather than on the ability to tap vast sums of money.

Superficially, governments and trade unions seem to be behind this drive towards wage equalization from country to country. This, however, is not the whole truth. The drive is also spurred by the big international companies, sometimes without their conscious awareness. It happens not necessarily because they are progressive and want things to develop this way, but simply because as they transfer people, they transfer values and demands as well as ideas and technology.

Thus, in the so-called capitalist system, it is the big companies whose success prompts social and economic demands from employees. The trade unions realize that the only way to participate in this development and promote it or steer it is to work towards internationalization themselves. Governments, by contrast, are last in line. They represent what are fondly termed "national interests" and supervise that portion of a company's activities located within a particular set of boundaries. As a result, because companies are mobile while governments are not, the governments have sometimes been left standing.

In some countries, success in the drive for equalization already means nobody needs to go without food or housing for lack of work. Physically we have achieved enough so that business can turn its attention to the mental and psychological elements of work. Work that means something to the worker has become one of the prerequisites for mental and physical well-being. Every individual should have the right to a job, and industry must make every effort to provide *good* work for those who are prepared to work.

But what does "right to employment" mean in a society where you can earn almost as much in unemployment pay as you can from a full working week? In some countries it seems as though the right to choose between working and not working has already begun to establish itself. In spite of practical differences between different regions and nations, the attitude to work is changing radically worldwide in these circumstances. Is it enough for work to be available? Or must the work be available near the worker's home? Should the worker be entitled to choose the job? Many people who still have heavy, arduous jobs are unwilling to pay for others to do exactly the kind of work they want—or no work at all. If a vote were taken on whether everybody, without exception, should be entitled to choose his or her work, most people would probably vote no, realizing that this is an attractive but unrealistic idea. For the time being, most of us know we must be prepared to choose among the jobs available.

## CREATING WORK

There are those who believe that in the next few decades work will become a privilege. Increased mechanization and automation are already reducing the demand for labor. This is a trend that deserves careful attention, for it could go awry. In the United States, for example, the Congressional Office of Technology Assessment reported in 1976 that automation in rapid transit systems brought few benefits and many problems. The expected savings for people operating trains were offset by the increase in (higher paid) people to repair breakdowns in the automated systems—an inadvertent kind of job creation. But jobs for both groups were less rewarding as a result. The train operators, a diminishing breed, felt their jobs had become the "residue of functions that equipment engineers have found technically or economically impractical to automate."*

The maintenance people, trapped in high-pressure, thankless jobs, suffered frustrations because they could seldom duplicate in

---

* "US Report Hits at Rapid Transit Automation," *New Scientist,* 15 July 1976.

their repair shops the conditions of normal operation. Reliability on one system went from 4000 hours between failures on fifteen-year-old cars to only 420 hours on the new automated cars. The analysis did not take into account the frustration of passengers waiting for cars that had broken down, or the essential shift in job focus from public service to equipment repair.

To my mind, the insistent demands for "work content" are a healthy sign. It is hard to arouse enthusiasm in routine occupations of diminishing importance to society. The demand to create better work may force us to reorganize production in ways that are also more efficient in the long run. An increasingly skilled and well-paid labor force can be a great asset. The difference between an apathetic worker and one with enthusiasm can have far-reaching results in quality as well as quantity of production.

Much of our discussion of work seems to center on production. When we talk about participation or meaningful (or meaningless) work we tend to see a blue-collar worker standing at his machine. But in truth, there is equal danger of dehumanization in white-collar jobs.

So far we have made little effort to create meaningful jobs— or to stop impoverishing those that are presently meaningful to their incumbents—in white-collar occupations. Just as humans have come to serve the machine in production, so too have attempts to make administration more effective failed; so far, most of the things done in offices in the name of "efficiency" simply narrow the work content and debase the human contribution. Public services at present are mainly "repair work." Dismal living conditions wear people down. So do child-care difficulties, commuting long distances, unrewarding jobs, and dealing with soulless bureaucracy. Better physical and social architecture could prevent some of these problems and reduce some of the need for social repair costs such as nursing and rehabilitation. At the same time, the jobs of those doing the repair work could become less erosive.

It may come to pass in the future that the public and private sectors will find themselves competing in the job-creation business —a competition that would ultimately be won on the basis of the quality, not quantity, of jobs created. So far, most of the effective

*A shop-floor planning meeting.*

steps in the direction of improving work, if not creating it, have been in the blue-collar area. Instead of pouring more money into so-called "production technology," some companies are beginning to spend it to adjust production to the demands of the people doing the work. It is possible to devise new solutions to combine rational technological systems with greater freedom for human choice.

Industrial work doesn't need to be tedious and dirty. But this is often the result when the only criterion is year-to-year economic growth measured in traditional terms. It requires a longer time scale to justify the technical development required to change systems towards more human satisfaction. Once this is achieved and human demands have been acknowledged, though, new possibilities open up that may be more effective in the longer term.

Every organization is different, so no single blueprint can be useful for everyone. Every individual is different, too—and that, in itself, is perhaps the single most important guideline for any organization that wants to create meaningful new jobs.

The rationale for job creation and job improvement is quite simple. Companies are already paying for anything from 20 percent to 50 percent more people than they need. If they accept this "cost" as part of their civic responsibility and the normal cost of doing business, they can then begin to make the "extra" work more meaningful, along with the so-called "essential" jobs." In many cases, this calls for an initial upheaval that is difficult for the managers in an organization, but the change in attitude is necessary to do business successfully under the conditions we are already experiencing today.

Volvo's experience in developing new work patterns at Kalmar, Torslanda, and other factories has given rise to a few rules of thumb that have been helpful:

- *The more that employes are involved in planning changes, the more successful the changes are likely to be.*

- *The highest probability of success seems to be associated with the change from line working to group working.*

- *It is vital that the individuals in a group have absolute choice over their method of work—some will want to keep on doing a single task all their working lives. There is no harm, and often some benefit to the organization, when they do so—so long as they do it by choice.*

- *"Job-creation" seems most effective when it takes the form of a few "extra' members in each group, beyond the technical minimum requirements. This gives room for flexible working hours, job-sharing (such as the situation in some of our plants where three young mothers share two jobs, or other arrangements of their own choice). This can be one of the most effective areas in which to create new jobs—and the payoff comes in increased customer satisfaction as well as job satisfaction.*

- *The one attitude that seems to mark the difference between success and failure is the one that says: "Employees are adults."*

On a larger scale, to evolve in this direction we must learn to design work to suit the labor available. And only the labor available can tell us what suits it. In national and corporate terms, this usually means changing our factories and offices and being willing to put them where people want to live. We must redefine production technology and administration technology, so the machines and materials are really serving the employees rather than vice versa.

In individual terms, changing work means learning to define the kind of job one wants, getting the training necessary to do it, and making sound decisions between the ideal of having it available right here and the practical limitation that some kinds of work will still require commuting or moving. As much as possible, within the realities of industrial and administrative life, people should be able to influence the design of their own jobs.

Without such changes, our present industrial and social problems can only grow worse. With new attention to the needs of the people who make up every organization, we should look forward to a new era of industrial growth and social progress.

## LEADERSHIP AND PARTICIPATION

A good deal of debate these days focuses on the qualities of leadership. Too often, colorful, strong leaders are thought to challenge or undermine the concepts of democracy, participation, and the delegation of decision making. The appropriate role of the leader is often described as that of compromiser, weighing various interests on a balance and then mapping out a middle-of-the-road route, using analysis and pragmatism to get there.

This concept of leaders and leadership is incorrect and dangerous. Analysis is and always will be helpful in all leadership tasks. To improve the quality of our decisions we naturally need more insight into the nature of a problem. However, in reality, analysis never leads to change. It never moves lines of demarcation. It draws up no new front lines. The only attributes of leadership that can result in essential changes are feeling and conviction. Risk taking is related to the unknown, and it is risk taking that is a hallmark in the development of a leader. Even if leadership demands a continuing ability to compromise, this should not lead to the disappearance of the most essential ingredient for development: courage.

Any group has within it creative capabilities. It is the collective effort of the group that can achieve results, sometimes of mountain-moving proportions. However, to awaken these capabilities, a group needs some force to wave the flag and rally mutual support for a common goal. This, as I see it, is the catalytic or triggering aspect of leadership.

People in a democracy are wary of leaders. This is natural, and a healthy reaction. Normally we are afraid of leaders only until we have seen them and have decided whether or not we like what we see. Familiarity gives us some opportunity for acceptance or rejection.

Of the leaders in our time, John F. Kennedy provides a fascinating example. No one can argue that his election as president of the United States was not the result of the democratic process. No one would contend that Kennedy, after he became president, made himself aloof from democratic institutions or public contact. Nor could anyone deny that John F. Kennedy exerted strong leadership in all matters, no matter what opinions might be about specific decisions.

Kennedy brought very strong personal judgments to the exercise of his office, and these values left few people untouched. When he died, the reactions of people all over the world indicated an absolute and very rare personal involvement. He left few people indifferent, whether they liked him or not.

People need the kind of personal sense of structure, and sheer leadership that Kennedy represented. He demonstrated that the ability to make firm decisions is not contrary to the democratic system, and indeed is necessary if democracy is not to fade into a colorless compromise that would eventually leave people indifferent to human conditions and issues outside their own circle.

On a smaller scale, exactly the same type of leadership is needed within every organization and every company.

When the role of a leader is debated, there is a tendency these days for people to assume that more participation and greater influence by the individual will erode leadership and demand less of leaders. I think this view is wrong. A leader must never mistake co-participation for laxity. There is a vital difference between positive delegation and the inability of a weak leader to accept his responsibilities. Real delegation requires that the leader transfer his influence, tolerance, and generosity to the delegate, and thus he must have the personal strength to afford sharing these properties. To open up an issue, to find enough courage to debate it, and to develop the strength to resolve a conflict—these are the qualities of true leadership. The weak are incapable of delegating and have every reason to fear sharing their power. The strong have the self-confidence that makes delegation possible and easy.

It is easier and more natural to accept new responsibilities if the role of the leader is clearly defined and if the final result expected from the leader is stipulated. A vaguely defined role coupled to a shifting debate makes the job harder, and hinders real democracy.

I cannot speak directly for the employees at Volvo. I can only observe how I see them behave in different situations. Since 1971, I have been able to see how union leaders conduct themselves as members of the company's board of directors. I have watched as they learned how the board functions and what kind of matters it handles. They now understand that the board is not engaged in activities that imply a conspiracy against the workers. These employee representatives have thus been able to negate many of the false myths about the board—perhaps one of the best and soundest results of broadening the board representation.

On our Corporate Works Council, I have, since 1971, seen employees discussing Volvo's entire strategy and planning operations. They have done this with an openness which clearly demonstrates the sense of responsibility they feel not only toward those employees they represent, but also toward the company as a whole. Politics and tactics are understandably part of their work. However, if one accepts the motives behind the tactics, one finds a strong desire for cooperation, based on the assumption that the workers have many interests in common with other stakeholders in the company. Furthermore, these people have a clear understanding of the leader's role. They respect leadership, and they understand the need to maintain the decision power within any type of organization, even if they must sometimes query the content and limits of leadership because of political convictions.

In another era, leadership was maintained by disciplinary rules and punishment. A company's management could apply sanctions or even fire people, ruling labor by decrees and notices. "Spitting prohibited," management could say. That time is long gone, and we can be thankful for the progress. Today we must grow closer to labor, no matter how busy management is.

Leadership is giving support, explanations, and interpreting information so employees can understand it. Leadership is developing consensus. Leadership is sometimes the ability to say "stop," to draw a line, to take the heat out of a conflict, to conclude a debate and get down to negotiations. Leadership is having the courage to put a stake on an idea, and risk making mistakes. Leadership is being able to draw new boundaries, beyond the existing limits of ideas and activities. Only through this kind of leadership can we keep our institutions from drifting aimlessly, to no purpose.

It sometimes scares me that what we do in Volvo is presented to others as an innovation, because this demonstrates, after all, how little has been done in work organization. Companies spend almost endless hours trying to provide change, incentive, interest, involvement, and motivation for top executives, yet almost no time is spent in looking at the rest of the work force in the same way. Until now, managers have not found it necessary. We are still in the era that Adam Smith described so many years ago, where "a worker gives up his ease, his liberty and his happiness when he goes into industry."

If we can give the worker back his ease, his liberty and his happiness, or at least provide conditions under which he can find them for himself, I believe we will come closer to a healthy, human, "postindustrial" society.